Springer Tracts in Civil Engineering

Series Editors

Sheng-Hong Chen, School of Water Resources and Hydropower Engineering, Wuhan University, Wuhan, China

Marco di Prisco, Politecnico di Milano, Milano, Italy

Ioannis Vayas, Institute of Steel Structures, National Technical University of Athens, Athens, Greece

Springer Tracts in Civil Engineering (STCE) publishes the latest developments in Civil Engineering - quickly, informally and in top quality. The series scope includes monographs, professional books, graduate textbooks and edited volumes, as well as outstanding PhD theses. Its goal is to cover all the main branches of civil engineering, both theoretical and applied, including:

- Construction and Structural Mechanics
- Building Materials
- Concrete, Steel and Timber Structures
- Geotechnical Engineering
- Earthquake Engineering
- Coastal Engineering; Ocean and Offshore Engineering
- Hydraulics, Hydrology and Water Resources Engineering
- Environmental Engineering and Sustainability
- Structural Health and Monitoring
- Surveying and Geographical Information Systems
- Heating, Ventilation and Air Conditioning (HVAC)
- Transportation and Traffic
- Risk Analysis
- Safety and Security

Indexed by Scopus

To submit a proposal or request further information, please contact:
Pierpaolo Riva at Pierpaolo.Riva@springer.com (Europe and Americas) Wayne Hu at wayne.hu@springer.com (China)

Guanghui Xu · Dongsheng Wang

Introduction to Intelligent Construction Technology of Transportation Infrastructure

Translated by George K. Chang, Hao Wang, Fei Wang, Shihui Shen, António Gomes Correia, Soheil Nazarian

 Springer

Guanghui Xu
Harbin Institute of Technology
Harbin, Heilongjiang, China

Dongsheng Wang
Harbin Institute of Technology
Harbin, Heilongjiang, China

Translated by
George K. Chang
The Transtec Group, Inc.
Texas, TX, USA

Hao Wang
Rutgers, The State University of New Jersey
Piscataway, NJ, USA

Fei Wang
Tarleton State University
Stephenville, TX, USA

Shihui Shen
Pennsylvania State University
Altoona, PA, USA

António Gomes Correia (iD)
University of Minho, Campus de Azurem
Guimaraes, Portugal

Soheil Nazarian
College of Engineering
The University of Texas at El Paso
El Paso, TX, USA

ISSN 2366-259X   ISSN 2366-2603   (electronic)
Springer Tracts in Civil Engineering
ISBN 978-3-031-13435-7   ISBN 978-3-031-13433-3   (eBook)
https://doi.org/10.1007/978-3-031-13433-3

This Springer imprint is published by the registered company Springer Nature Switzerland AG
The registered company address is: Gewerbestrasse 11, 6330 Cham, Switzerland

Foreword to the English Edition

This English edition is a translation of the Chinese version of "Introduction to Intelligent Construction for Transportation Infrastructure," with the Chinese Title "交通基础设施智能建设技术导论". Professor Guanghui Xu initiated the Chinese book series for Intelligent Construction Technologies (ICT), starting with this book. The Chinese edition was published by China Railway Publishing House (CRPH) Co., Ltd. in November 2020. In March 2021, CRPH and Springer Nature Switzerland AG signed a license agreement to publish this English version in the book series "Springer Tracts in Civil Engineering". The English edition aims to reach out to global readers with the ultimate goal of training and cultivating the current and next-generation intelligent construction technologies workforce.

In March 2021, the Executive Committee (EC) of the International Society for Intelligent Construction (ISIC) decided to assist CRPH in translating the Chinese ICT book series into English editions. The EC includes Dr. George K. Chang (Transtec Group, USA; President of ISIC), Prof. António Gomes Correia (University of Minho, Portugal; Vice President of ISIC), Prof. Guanghui Xu (Harbin Institute of Technology and Southwest Jiaotong University, China; Vice President of ISIC, also the primary author of the Chinese edition), and Prof. Soheil Nazarian (University of Texas at El Paso, USA; Vice President of ISIC).

The EC established the translation plan and invited several US scholars with English and Chinese language skills to accomplish the English translation work. The translators include Dr. George K. Chang, Prof. Hao Wang (Rutgers University, USA), Prof. Fei Wang (Tarleton State University, USA), and Prof. Shihui Shen (Pennsylvania State University—Altoona, USA).

The English translation work was divided into: Chapters 1 and 7 by Dr. George K. Chang; Chapters 2 and 6 by Prof. Hao Wang; Chapter 3 by Prof. Fei Wang; and Chaps. 4 and 5 by Prof. Shihui Shen. The reviewers for the English version include Prof. Antonio G. Correia and Prof. Soheil Nazarian.

The EC acknowledges the efforts of the above English translators and reviewers as well as the coordination efforts between CRPH and Springer to make this English edition possible. EC encourages any feedback from readers to be sent to GKChang@TheTranstecGroup.com.

International Society for Intelligent Construction (ISIC)

Preface to the First Chinese Edition

The intelligent construction of transportation infrastructure is a new field as the product of multidisciplinary cross-integration. It is the concrete manifestation of a new generation of information technology in engineering construction, which will change the technical system of the traditional construction industry. Since it is a new field, the basic concept, technical composition, and application of intelligent construction are still inconclusive. Because of this, how to build the technical architecture of intelligent construction is very critical, involving issues such as technical composition.

Based on years of practical experience in intelligent construction monitoring technology and the latest research results in related fields, the authors conclude that intelligent construction should have the essential characteristics of "perception, analysis, decision-making, and execution." Among them, the "perception" part involves electronic information technology, sensing technology, detection technology, and domain knowledge; "analysis and decision-making" involve artificial intelligence technology and domain knowledge; the "execution" involves control technology, domain knowledge, and engineering machinery. This book is the first volume of the "Frontiers of Transportation Infrastructure Intelligent Construction Technology" series. It focuses on the essential characteristics of intelligent construction, mainly expounds on the relevant technologies of intelligent construction of transportation infrastructure, and involves most of the content in the series. This book is an overview of intelligent construction technology.

Chapter 1 summarizes the technical characteristics and critical points of intelligent construction; firstly recognizes the essence of artificial intelligence and the method of realizing "intelligentization"; then establishes the basic structure of intelligent construction technology. Chapter 2 provides a set of systematic methods for mastering new technologies, gives the characteristics of five fundamental problems of the system, and introduces the technical characteristics of the road structure system, artificial intelligence system, intelligent construction system, and information platform. Chapter 3 analyzes intelligent algorithms or machine learning algorithms (such as neural networks, deep learning, and reinforced learning), restores a real artificial intelligence, and briefly describes the relationship between machine learning and engineering construction. Chapters 4 and 5 reveal the nature of big data and the Internet of Things (IoT) foundation behind the information platform, introduce several perception technologies in engineering, and discuss data security issues. Chapter 6 expounds on the feasibility and technical solutions of the application of various intelligent technologies in design, construction, and maintenance; discusses the application of virtual reality technology in engineering construction; takes intelligent compaction as an example to demonstrate the realization of "perception, analysis, decision-making, execution" steps. Chapter 7 briefly explains that improving engineering quality is the short-term goal of intelligent construction and discusses the essential characteristics and implementation methods of information highways and railways, intelligent highways, and intelligent railways. According to the essential characteristics of intelligent construction, the final chapter combines intelligent construction talent training and sorts out the required professional knowledge and the theoretical basis behind it.

Part of the content of this book has been taught several times to the undergraduates of Harbin Institute of Technology (including those majoring in road, bridge, traffic engineering, and traffic information control). This book has been revised several times according to the feedback information. It took more than three years to complete the manuscript. Even so, since intelligent construction is a new field and is still in its infancy, how to grasp the technical composition of intelligent construction is still in the exploratory stage. This book draws the attention of more technical personnel in the field and jointly promotes intelligent construction technology's sound development.

The founders of the International Society for Intelligent Construction (ISIC), Dr. George K. Chang of the Transtec Group in the USA, Prof. Soheil Nazarian of the University of Texas in El Paso in the USA, and Prof. António Gomes Correia of the University of Minho in Portugal, participated in the overall planning of the series and this book. Dr. George K. Chang also provided some original content. They also participated in the reviews and edits of the English version.

Due to the limited knowledge of the authors, this book consists of their own thoughts, and some misses may be inevitable. Therefore, we urge readers to send critics and feedback to highx@163.com.

Guanghui Xu

Preface to the Second Chinese Edition

With the re-emergence of artificial intelligence (AI) technology, almost all fields are developing in the direction of intelligence, hoping to use AI technology to promote innovation in this field. This phenomenon can be collectively referred to as "AI+." Inspired by intelligent manufacturing, starting from 2016, we have been researching intelligent issues in the road and railway engineering field in Harbin, China. We have provided such materials to the undergraduate students in the School of Transportation Science and Engineering of Harbin Institute of Technology (including roads, bridges, traffic engineering, and traffic information control major). We have set up a course on intelligent construction to teach the application of multidisciplinary knowledge in the field of transportation infrastructure. We soon realized that the so-called AI+ is not a simple "AI+ field technology" but the integration of modern information technology and a specific field technology (professional technology), which is the technological upgrading of traditional industries. This requires us to break the traditional disciplinary boundaries and view and study related issues in this field from a broader perspective.

Since the first edition of this book was published in October 2020, it has sold out for almost a year, and some colleges and universities use it as a text book. To meet the needs of development, this revision rewrites and draws the structure and content of the book, deletes parts that are not closely related to intelligent construction, and adds new content to meet the needs of the rapid development of intelligent construction. This book can also be arranged in the order of "perception, analysis, decision-making, execution", the basic characteristics of intelligent construction (the teaching order can also be arranged in this way). It can also be taught by highlighting machine learning first and letting readers understand the essence of AI as soon as possible. This part of the content is placed in the front but does not affect the reading order. Because these four basic characteristics correspond to different disciplines, they exist independently.

Although there are various references to "AI+", the basic idea is the same—using machines to do work instead of humans. Since human beings do work along the basic sequence of perception→analysis→decision→execution, "perception, analysis, decision-making, and execution" have become the basic characteristics of

"AI+". "Perception technology" includes perception terminals based on microelectronics technology and perception methods based on physics. "Analytical technology" is based on data science, and machine learning is only a technical means of data analysis. "Decision-making technology" not only requires the intervention of AI technology, but it also needs more support from field technology. "Execution technology" mainly exists in various fields, but the common foundation is the control theory. Based on the above considerations, this book mainly introduces machine learning and data knowledge in "Analysis and Decision" and electronic terminal knowledge in "Perception". At the end of the book, it facilitates readers to grasp the technical system of "AI+" from a macro perspective. Starting from the basic characteristics of intelligent construction, it sorts out artificial intelligence, electronic technology, dynamics, big data, control technology, cloud computing, Internet of Things, the relationship between communication technology and domain (professional) technology, etc.

When this book is used as a university teaching material, the teaching can be arranged for a 32-hour curriculum. Don't hesitate to contact the publisher if you need electronic teaching material. At the same time, there will be more discussions and related materials for readers' reference on the International Society for Intelligent Construction (ISIC) expansion platform (www.isicshow.com in English, www.isicshow.cn in Chinese).

This revision has received extensive support from all walks of life in China and abroad. Many experts have put forward revision opinions from different angles, and we would like to express my heartfelt thanks. Limited by the author's knowledge, the book may still have some inaccuracies. Please send feedback to highx@163.com.

Guanghui Xu

ISIC's Endorsement

With modern science and technology development, the traditional transportation infrastructure construction industry is undergoing profound changes. The combination of modern information technology represented by artificial intelligence and the traditional construction industry has become an inevitable trend. In this context, intelligent construction is proposed to improve engineering construction quality and ultimately realize intelligent transportation infrastructure.

Transportation infrastructure construction belongs to the traditional civil engineering industry. Due to various reasons, many civil engineering technicians have limited knowledge of modern information technology, and even some terms are unclear. Therefore, we need to change the technical system of the traditional construction industry and update our knowledge. The cross-integration of different disciplines and majors has become inevitable. This is called "long-term integration." It is against this background that this series of books is proposed. The purpose is to break the boundaries of disciplines, integrate multidisciplinary knowledge, initially establish a technical system for intelligent construction, and promote the development of intelligent construction technology.

Currently, many "AI +," including intelligent construction, do not have a complete technical system. Different people and industries have different understandings of intelligent construction, which is an inevitable phenomenon in the development process. We have been researching this issue for many years. Based on practical experience, basic ideas for developing intelligent technology applications are summarized. One of the purposes of developing artificial intelligence is to allow computers to do work instead of human brains. It is necessary to analyze the fundamental processes and characteristics of people doing work. According to our practical experience in road and railway engineering intelligent construction monitoring technology, it is feasible to implement intelligent construction by following the technical route of "perception, analysis, decision-making, and execution." Therefore, the book series uses the transportation infrastructure (highways, railways, airport pavements, and urban roads) as the background to carry out the overall concept and implementation of intelligent construction around the four essential characteristics of "perception, analysis, decision-making, and execution." This book series, taking into account theory and practice, would help engineers and technicians in related fields understand and master new technologies in engineering construction.

The book series uses engineering construction as the background, combined with the applications, and discusses the theory, technology, and engineering application involved in the aspects of "perception", "analysis", "decision-making", and "execution." The books include (1) Introduction to Intelligent Construction of Transportation Infrastructure; (2) Foundations of Perception Terminals: Information Technology in Engineering; (3) Foundations of Perception Methods: One-Dimensional Dynamics and Applications in Engineering; (4) Foundation of Machine Analysis and Decision-Making: Entering Machine Learning; (5) The Weapon of Engineering Quality: Perception and Data; (6) Assistant to Execution: Control Technology in Engineering; and (7) Pioneer of Intelligent Construction: Intelligent Compaction. Series (1) is a framework that is a general discussion of "perception, analysis, decision-making, and execution", covering the main content of intelligent construction technology. Series (2) takes the perception terminal as the background and introduces the related electronic information technology. Series (3) introduces the vital basis of perception methods—one-dimensional dynamics and application, which is the basis of various non-destructive testing techniques. Series (4) introduces how to use machines to replace human brains for data analysis and decision-making, which is machine learning as an upgrade of traditional data analysis techniques. Series (5) introduces perception technology and data analysis technology (data science) together. It is mainly because perception is the primary technical means of data acquisition, data is the result of perception, and data analysis is also the basis of decision-making and a tool for mining information and knowledge in data. Series (6) introduces the assistant of "execution"—control technology, which is one of the common foundations in various fields. Series (7) is the application case study in intelligent compaction, which describes how to "perceive, analyze, decide, and execute."

The intelligent construction of transportation infrastructure is one of the frontier issues in the construction industry. As for new areas, there are still many unknown to explore. A journey of thousands of miles begins with a single step, and laying a solid foundation is the key. Knowing and mastering the necessary multidisciplinary knowledge is a tool and a weapon in the path of exploration. China Railway Publishing House (CRPH), in conjunction with the International Society for Intelligent Construction (ISIC, www.is-ic.org), organized and compiled this frontier series of intelligent construction technology for transportation infrastructure for the first time in the world. This book series is indispensable and timely and can help construction engineers and technicians in related fields broaden their knowledge, inspire ideas, and support the popularization of new concepts and technologies in intelligent construction.

The series is published in Chinese edition (by CRPH) and English edition (by Springer Nature). The authors and translators are from China, the USA, and the European Union. They are all members of the International Society for Intelligent Construction. They have different professional backgrounds and rich practical experience. They understand the current situation of technical personnel in the engineering construction industry, making the writing more targeted and readable. The series has the following characteristics.

(1) New fields. The intelligent construction of transportation infrastructure belongs to a new field and is the product of multidisciplinary intersection. The series of books is compiled for the first time in the world, which can help technical personnel in engineering construction to understand and master new knowledge and technologies in the information age.

(2) Innovation. Intelligent construction is an interdisciplinary subject, and its technical composition is still inconclusive. From the perspective of engineering construction, the series of books explores the establishment of the basic structure of intelligent construction technology, extracts relevant knowledge from multiple disciplines, and carries out systematic writing and integrated innovation.

(3) Practicality. The series of books is based on engineering construction and revolves around "perception, analysis, decision-making, and execution." The engineering cases involved are all derived from practical applications and have been validated in the fields.

(4) Readability. The authors avoid using obscure professional language but use popular language to explain professional issues, which is convenient for technical personnel in engineering construction to read.

Combining the characteristics of engineering construction, the series of books expound on the related technologies of intelligent construction from multiple technical perspectives. The book series provides a set of professional-level introductory materials for technical personnel in related fields such as highways, railways, airports, and urban roads to understand, learn and master new intelligent construction technologies. Readers with college knowledge can understand most of the content. At the same time, it will also play a role in broadening horizons and inspiring research ideas for undergraduates and postgraduates in related majors in colleges and universities. It can also be used as a textbook.

Although the series is based on engineering construction, its fundamental ideas, technical routes, and various technologies are applied to other fields as long as the application background is modified accordingly.

International Society for Intelligent Construction (ISIC)

Contents

1 What is Intelligent Construction? 1
  1.1 Construction Characteristics of Transportation Infrastructure 1
  1.2 Life Cycle and Construction Quality 6
  1.3 Engineering Construction Needs the Help of Modern
      Information Technology 7
  1.4 A Preliminary Understanding of AI 9
  1.5 Understand Intelligent Construction Technologies 16
  1.6 The Keys to Implement Intelligent Construction Technologies 21
  1.7 Framework of Intelligent Construction Technology 24

2 Systematic Approach to Representing Objects 35
  2.1 Definition of the System 35
  2.2 Basic System Problems 41
  2.3 The Black Box Method 43
  2.4 Road Structure System 46
  2.5 Artificial Intelligence System 48
  2.6 Intelligent Construction Technology System 49
  2.7 Information Platform System 52

3 Empowering Machines to Learn 57
  3.1 A Computer Simulates Human Brains 57
  3.2 How Do Machines Learn 61
  3.3 Dissect an Intelligent Algorithm 66
  3.4 Deep Learning is not a Mystery 93
  3.5 Other Algorithms That Can't Be Ignored 101
  3.6 Return to Reason: Look at AI Objectively 110
  3.7 Machine Learning in Engineering Projects 112

4 Entering a Data Era 115
  4.1 Embracing the Data Era 115
  4.2 Big Data Versus Small Data 118
  4.3 Digitization of Engineering Information 120

4.4 Causality and Correlation 121
4.5 Knowledge and Data 123
4.6 Data Analysis Methods 124
4.7 Data Security 130

5 Understanding Perception Technology 133
5.1 How is the Data Obtained? 133
5.2 The Boom in Perception Technology 135
5.3 The Fundamentals of the Perception Technology 138
5.4 Ways to Learn Perception Techniques 141
5.5 Sensoring Devices in Engineering 142
5.6 Automatic Sensing and IntelliSense 145
5.7 Internet of Things in Engineering 146

6 Intelligent Engineering Construction 153
6.1 The Dawn of Intelligent Construction 153
6.2 The Feasibility of Intelligent Engineering Design 156
6.3 The Rise of Intelligent Construction 167
6.4 Intelligent Maintenance has Started 189
6.5 Intelligent Management Technology 196
6.6 Virtual Construction Has Great Potential 199
6.7 Intelligent Construction Example: Intelligent Compaction 203
6.8 The Risks of Intelligent Construction 213

7 The Road to the Future 217
7.1 Quality is the Key 217
7.2 Informatization of Roads Depends on the Cost 218
7.3 The Path Forward is Innovation 220
7.4 The Future of ICT is at Our Doorstep 221

References 227

About the Authors and Translators

Authors of the Chinese Edition

Guanghui Xu Ph.D. (https://orcid.org/0000-0002-4004-2374) worked in railway design and research institutes in his early days in China. He later joined Harbin Institute of Technology and other Chinese universities to engage in scientific research and teaching. His research interests include road and railway engineering dynamics theory, testing and information analysis technology, and intelligent technology applications. Professor Xu is a co-founder of ISIC. He has always adhered to the principle of paying equal attention to theoretical and applied research and practical results, organized and led a multidisciplinary scientific research team, and carried out long-term independent research and development of intelligent construction monitoring technology. Professor Xu also formed a series of research results with intellectual property rights. In 2011 and 2017, he presided over the compilation of China's first industry construction standards and product standards for intelligent compaction (IC) technology. He also published an IC monograph in Science Press and China Railway Press.

Dongsheng Wang Ph.D., is a Road and Railway Engineering Department Professor at Harbin Institute of Technology, China. Professor Wang is also an ISIC Steering Committee member and a member of the Heilongjiang Youth Federation. His main research areas are pavement mechanics and structural design theory, pavement materials, and constitutive relations. He has led over nine projects, including the National Natural Science Foundation of China, and won the second prize of the Jilin Province Science and Technology Progress Award and the first prize of the China Highway Society Science and Technology Award. Professor Wang has participated in the compilation of many industry standards. He was selected into the Harbin Institute of Technology's top-notch teaching talents program. He has won the school's gold-medal teaching teacher, the first prize for teaching excellence, the first prize of the National University Young Teacher Teaching Competition, and the person in charge of the school's graduate course "Pavement Mechanics and Analysis Methods."

Translators of the English Edition

George K. Chang Ph.D., P.E. (https://orcid.org/0000-0002-4945-8827) is the director of research of the Transtec Group, USA. Dr. Chang is a co-founder and president of ISIC. He is a world expert on pavement smoothness and intelligent compaction/construction technologies. His research, teaching, specification development, and software tools have helped make significant technological advancements in the above fields. He has been leading the US national deployment effort intelligent compaction since 2007. Dr. Chang has been the chairman of the Road Profile Users' Group (RPUG), TRB AFD90/AKP50 Pavement Surface Properties and Vehicle Interaction committee, and ASTM E17.31 Profile Measurement subcommittee. He is an Emeritus member of the TRB AFD90/AKP50 committee. Dr. Chang received many industry awards, including the US Highway Intelligent Compaction Innovation Technology Award, ASTM International awards, and a TRB AKP50 lifetime emeritus member.

Hao Wang is Associate Professor and Graduate Director of Civil and Environmental Engineering at Rutgers University, USA. Professor Wang's general research is on the sustainable, intelligent, and resilient built environment. His recent research focuses on (1) multi-scale modeling and characterization of pavement material; (2) development of multi-functional infrastructure material; (3) sustainable and innovative pavement in highways and airfields; (4) life-cycle analysis and assessment and pavement management system. His research has been sponsored by many US Federal, State, and local transportation agencies. He has authored over 150 journal and conference publications in pavement engineering and infrastructure material.

Fei Wang is Assistant Professor in Civil Engineering at the Tarleton State University, USA, a professional engineer in Missouri and Kanas, USA. Professor Fei Wang's research interests include Geo-Hazards Monitoring, Evaluation, and Mitigation, Soil-Structure Interaction, Performance of Buried Culverts, Tunneling, AI-Based Technology for Civil Engineering, Numerical Modeling in Geomechanics, Geosynthetic Reinforcement, Sustainable Material in Civil Engineering.

Shihui Shen is a Professor of Engineering at the Pennsylvania State University—Altoona, USA. Professor Shen is a professional engineer in Pennsylvania and a fellow American Society of Civil Engineers. She is also an affiliated graduate faculty member at Penn State College's Department of Civil and Environmental Engineering. She received her doctoral degree at the University of Illinois at Urbana—Champaign, USA. Dr. Shen studies pavement materials and structures for better durability, energy efficiency, and sustainability. In addition to her ongoing interest in civil material characterization and innovation, Dr. Shen has recently worked on structural health monitoring and modeling for pavements using wireless sensors and data-driven approaches. Her research has been funded by the US state DOTs, NSF, NCHRP, USDOT University Transportation Center, FAA, US Army Corps of Engineering, and private sectors. Dr. Shen has published more than one hundred peer-reviewed technical papers and reports and serves on many research committees, journal editorial boards, and project panels for the professional community.

António Gomes Correia (https://orcid.org/0000-0002-0103-2579) is Professor at the University of Minho, Portugal. Professor Correia is an ISIC co-founder and vice president. He has been engaged in research, teaching, and consulting geotechnical and pavement engineering for over 35 years. His scientific research includes transportation geotechnical engineering, especially soil and road geotechnical material performance, modeling, compaction, soil improvement, foundation, geotechnical engineering design, management, etc., with more than 360 technical papers and 240 reports on these topics. He has also supervised 116 graduate students, including 30 doctoral students. From 1998 to 2001, Dr. Gomes Correia served as the chairman of the ISSMGE—European Technical Committee—ETC 11 on Geotechnical Aspects in Design and Construction of Pavements and Rail Track. In 2001, he became the chairman of the ISSMGE Technical Committee TC 3 on Geotechnics for Pavements, renamed in 2009 as TC 202—Transportation Geotechnics. He was the Chairman of TC 202 until 2013 and after a member of the Executive Group. From 2009 to 2013, Dr. Gomes Correia also served as a European Member of the Technical Oversight Committee (TOC) of ISSMGE. From 2004 to 2008, He was also President of the Portuguese Geotechnical Society. Professor Correia is a well-known European intelligent compaction expert and the editor-in-chief of the European Union's intelligent compaction standards. Professor Correia also demonstrated his leadership in starting a new international technical journal, *Journal of Transportation Geotechnics*, launched in September 2013 by Elsevier's Engineering Journals. He is currently the Editor-in-Chief of the *Journal of Transportation Geotechnics* and serves on the Editorial Boards of other National and International Journals.

Soheil Nazarian is Professor at the University of Texas at El Paso (UTEP) USA, the McKentas-Morkinson director of civil engineering, and the Transportation Infrastructure System Center of the school. Professor Nazarian is an ISIC co-founder and subcommittee leader. Dr. Nazarian has more than 25 years of experience in materials and non-destructive testing related to geotechnical and transportation infrastructure. He has been the PI and co-PI of more than 100 research projects funded by US federal and state agencies such as the Texas Department of Transportation, Federal Highway Administration, and Strategic Highway Research Program. Throughout his career at UTEP, he has supervised over 80 Master and Ph.D. students. Many work for the Texas Department of Transportation and other private and public transportation entities.

Chapter 1
What is Intelligent Construction?

Abstract This chapter first introduces the essential characteristics of transportation infrastructure, analyzes the relationship between modern information technology and engineering construction, and then introduces the origin and purpose of artificial intelligence from the perspective of the Turing test. Based on analyzing what human beings do, the essential characteristics of intelligent technology in various fields—"perception, analysis, decision-making, and execution"—P.A.D.E. are put forward, and the basic meaning of intelligent construction and the key technologies to be developed at present are given. Finally, taking highway engineering as an example, the main contents of the application of intelligent technology in the design, construction, and maintenance stages are analyzed, and the basic structure of intelligent construction technology of transportation infrastructure is established on this basis.

1.1 Construction Characteristics of Transportation Infrastructure

The intelligent construction mentioned in this book is mainly for transportation infrastructure. Practical experience shows many common technologies in all "AI+" fields, which is the book's focus. The intelligent construction of transportation infrastructure is an example of the application of intelligent technology.

In order to grasp this new technology, it is necessary first to understand the relevant knowledge about infrastructure construction. This is the technology of individuality in "AI+".

(1) Infrastructure

Infrastructure provides services for the transportation of goods and people. More generically, infrastructure includes public facilities (such as roads, railways, airports, telecommunications, water, electricity, and gas) and social facilities (such as education, technology, medical care, sports, and culture). Therefore, infrastructure is the foundation of the national economy. Its significance is self-evident. Figure 1.1

© China Railway Publishing House Co., Ltd. 2023
G. Xu and D. Wang, *Introduction to Intelligent Construction Technology of Transportation Infrastructure*, Springer Tracts in Civil Engineering, https://doi.org/10.1007/978-3-031-13433-3_1

Fig. 1.1 Examples of instructure

shows examples of infrastructure, including highways, city roads, electricity, telecommunication, bridges, rails, and ports.

The transportation infrastructure is an essential part of the infrastructure, the basis of the "travel" in clothing, food, housing, transportation, and the passage of transportation. It is closely related to people's lives and is the focus of our attention.

Transportation plays a role as a link in the national economy and is divided into land transportation, air transportation, and water (marine and river) transportation. The infrastructure corresponding to land transportation can be summarized into three parts: roads, bridges, and tunnels. The "roads" mainly refer to railways and highways, as well as urban roads. The infrastructure for air transportation is mainly airports. The corresponding infrastructure for water transportation is the port. The rest of the narrative mainly focuses on roads and railways (that is, roads and railway projects), and airports and ports can also be attributed to a part of "roads." Fig. 1.2 shows some examples.

Fig. 1.2 Transportation instructure

The construction of transportation infrastructure is both familiar and unfamiliar. The so-called familiarity gives people the impression that it is commonly referred to as "repairing railways and roads." Everyone can understand these construction behaviors. The so-called unfamiliar means that the construction of transportation infrastructure includes related technologies in many disciplines, which is not as simple as everyone thinks.

(2) **Transportation infrastructure**

Traditional transport infrastructure mainly refers to roads and railways. This kind of linear structure stretches hundreds or even thousands of kilometers. It consists of three parts: "road, bridge, and tunnel". Bridges and tunnels are control projects under construction and belong to "point" structures; lines (referred to as roads) are the main projects and belong to linear structures. The three structures are different regarding raw materials and appearance, and the design and construction methods are also very different. These differences lead to their different construction characteristics.

1. **Bridges**

Bridges are man-made structures composed of reinforced concrete or steel. The design and construction are carried out by structural engineering. The bridge's lower foundation is generally mostly piles foundation; the upper part is the structural parts, main beams, and slabs, which are generally constructed by the precast and assembly method. Figure 1.3 shows an example of a bridge construction site.

The design and construction technology of bridges are relatively advanced. As long as the relevant technical standards are strictly implemented, a good structure that meets the requirements can be obtained.

2. Tunnels

The tunnel is an excavation structure (excavated mountain or underground rock and soil) composed of rocks, soils, and a reinforced concrete lining structure (supporting structure). Its design and construction are generally carried out by geotechnical engineering. The key to the tunnel is construction, mainly dealing with surrounding rock problems (such as collapse and water leakage). With the maturity of large-scale construction machinery (such as shield machines) and advanced forecasting technology, the construction difficulty has been reduced, and safety has been further improved. The practice has proved that as long as the construction process is

Fig. 1.3 Bridge construction

Fig. 1.4 Tunnel construction

strictly managed and scientifically constructed, there will be no significant problems. Figure 1.4 shows an example of a tunnel construction site.

3. Road

If bridges and tunnels are considered point-type structures, roads and railways are belt-type or linear structures that stretch for hundreds of kilometers. This kind of structure is different from bridges and tunnels. It is a geostructure ("dikes" and "grains") formed by filling and rolling various types of construction materials or excavation of mountains. The type structure dominates and can be called a filling project. This kind of structure has its unique design and construction method.

In terms of design, the "road" belongs to a linear structure, and its linear shape must satisfy the kinematics of the vehicle and match the surrounding environment. The structural form is an "inverted trapezoid" layered body composed of geotechnical materials, concrete, and asphalt mixtures. In terms of construction, materials need to be paved and rolled to form a filling structure. In terms of maintenance, it involves more various detection techniques and data analysis and evaluation techniques, which are multidisciplinary integration.

The linear infrastructure consists of the railway (subgrade and track) and highway (subgrade and pavement). The subgrade structure is the same, and they are all composed of rock and soil fillers. The superstructure of the railway is divided into ballast track (gravel ballast bed + sleeper + steel rail) and ballastless track (concrete slab + track slab + steel rail), as shown in Fig. 1.5.

For highways, the part above the subgrade structure is called the pavement structure, composed of "base layer + surface layer". The base layer is divided into a

Fig. 1.5 Railway construction with ballast and concrete/ballast-less bases

Fig. 1.6   Roadway subbases, concrete pavement, and asphalt pavement

Fig. 1.7   Quality control focuses on filling works

semi-rigid base layer and a flexible base layer. The surface layer is divided into a concrete surface layer and an asphalt surface layer, as shown in Fig. 1.6.

According to the above brief description of the basic characteristics of roads, bridges, and tunnels, in the transportation infrastructure, lines belong to the filling type, and the proportion is also large (except for high-speed railways in China), and their engineering quality is also high. Relatively the most difficult to control, as shown in Fig. 1.7.

Generally speaking, whether it is a highway or a railway, filling works such as roadbeds and pavements are difficult to grasp. The properties and types of fillers are far more complicated than concrete ones and are closely related to the environment (water and temperature). The construction quality is also difficult to fully control and has always been the "bottleneck" of construction technology. These factors also cause premature road and railways' premature damage. Therefore, special attention must be paid to reclaimed buildings such as "roads". It is also necessary to use "external forces" to promote the innovation of construction technology, help them improve engineering quality, and prolong service life, which is also the focus of intelligent construction at this stage.

1.2 Life Cycle and Construction Quality

Now more and more industries are accustomed to using anthropomorphic words to express a certain meaning. For transportation infrastructure, the concept of full life cycle is an anthropomorphic expression (also called health inspection, lifeline engineering, etc.), which is a very pictographic term and easy to understand.

The whole process from birth to death is called a life cycle. For example, a person lives 90 years old, then 90 years old is his life cycle, also called life span. Compared with people who have lived 60 years old, 90 years old is a long life. Of course, it must be healthy and happy to live! In fact, it is the performance of the high "quality" of the body.

For a product, the whole process from birth to obsolescence is its life cycle, also called service life. For example, the usable time of a car is generally 15–20 years, which is its life cycle or service life. If the car often has various problems during use and needs frequent maintenance, then it can be said that the quality of the car is very low. The quality of a car is closely related to the materials and manufacturing technology used. Transportation infrastructure is also a product with similar meanings.

Taking road engineering as an example, the whole process from construction to scrapping is the life cycle of the road. The life cycle is generally controlled by the structure of the road surface (compared to the roadbed, the road surface is more likely to be damaged). Therefore, the life cycle of a road project is actually the life cycle of the pavement, that is, the entire time (in years) of the road from construction → operation → maintenance → scrapping is the life cycle of the pavement and the service life of road structures, as shown in Fig. 1.8 shown.

If a road is maintained less frequently in its life cycle, its engineering quality is higher, and vice versa. The number of repairs in Fig. 1.8a is 2, and the number of repairs in Fig. 1.8b is 4. We would say that the former is in a higher quality state, and the latter is in a lower quality state.

Regarding engineering quality, there is no complete definition at present. The engineering quality includes the internal and the external portions. The internal quality includes the

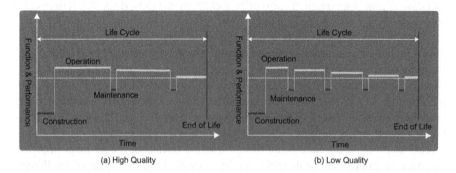

Fig. 1.8 Life cycle and construction quality

functions, durability, reliability, applicability, etc., of the project; the external quality includes the beauty of architectural art, engineering technology, the coordination with ecology, the integration of culture, and the service.

No matter what products, quality issues are critical. Transportation infrastructure construction aims to improve quality by systematically involving the life cycle. Intelligent construction technologies are current and emerging means to achieve this goal.

1.3 Engineering Construction Needs the Help of Modern Information Technology

It is not enough to rely solely on conventional civil engineering methods to improve the transportation infrastructure. It requires the intervention and integration of civil engineering with other disciplines. Information technology is such an example.

Modern information technology includes, but is not limited to, sensing and testing technology, computer technology, artificial intelligence, communication technology, network technology, automation technology, control technology, etc. It is the most active modern technology that improves productivity and construction quality.

Information technology for construction includes computer-aided design (CAD) to various management and control tools to improve efficiency and quality. Smartphones, intelligent laundry machines, and smart air conditioners are examples of information technology applications, among many others, in our daily lives.

The core of information technology is computer hardware and software. The computer chip is the hardware's core, and the programming algorithm is the software's core. Figure 1.9 shows the relationship between chip, circuit board, and devices (including computers).

At present, microcomputers (such as single-chip microcomputers) have been embedded in various devices (items), becoming the "brains" of the devices. In these devices, sensing technology, communication technology, and control technology are

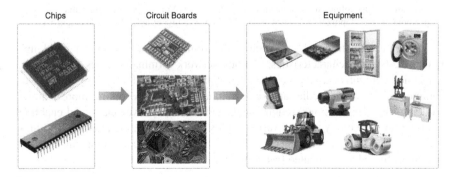

Fig. 1.9 From chip to the circuit board and devices

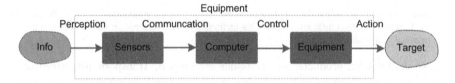

Fig. 1.10 Sensing, communication, and control in equipment

all indispensable components. The sensor is responsible for the perception of information, the communication is responsible for the transmission of information, and the control is to determine the behavior of the executing agency based on the information. Figure 1.10 shows this process.

Microelectronics and software technology are the keys to computers and support information technology. It is common to embed computer chips in various devices. Each chip contains hundreds of millions of micro-components, significantly improving the device's function and reducing size, weight, and power consumption.

> The chip is an integrated circuit with a massive number of micro-transistors. The manufacturing technology is complicated. In many cases, integrated circuits and chips are often used interchangeably. The chip is the core of the microprocessor unit (MPU), controlling computers and various electronic devices. Chips continue to be smaller but pack more circuits.

For software technology, it has shifted from being computer-centric to network-centric (cloud computing, cloud services). The interpenetration of software and integrated circuit design makes the chip become "hardened software", further consolidating the core position of the software. With the rapid development of software technology, many functions can be realized by software, and "hardware softening" has become a trend. The development of embedded software makes software technology go out of the traditional computer field, promotes the popularization of goods intelligence, and becomes one of the core technologies for advancing informatization.

> What is the basis of microelectronics technology and algorithms? The answer is physics and mathematics, and physics before the 1960s, mathematics is even more distant. The practice has proved that no matter how engineering technology develops and changes, its foundation is in essential disciplines such as physics and mathematics. Modern physics and mathematics promote the development of science and technology to a higher level. Let us wait and see.

Modern information technology develops around the Internet, and "networking" is one of its main characteristics. Various network terminal devices with single-chip microcomputers as the core are developing rapidly. Technology such as e-commerce, electronic media, and remote monitoring are becoming more mature, reducing users' professional requirements and economic investment, and enabling information technology to enter various fields. On this basis, the Internet of Things makes it possible to interconnect everything. Through Cloud computing, information can be processed and executed faster and better.

The development of information technology has also brought new opportunities to engineering construction. Various new technologies with information technology as the core are rapidly infiltrating and merging, changing traditional industries' technical systems. The "informatization" and "networking" of engineering construction have been realized, and the pace of "intelligence" is also accelerating.

> The term "informatization" was initially associated with "information industry" and "information society," and it originated from the book "On Information Industry" published by Japan in 1963. Since then, the term "informatization" has been widely used worldwide. People have explained the definition and connotation of "informatization" from multiple angles, but no standard definition has been formed. The current so-called informatization is another term for "computerization or digitization."

Information technology is a boost that can help the traditional construction industry update its technology. But learning new technology does not mean giving up traditional technology. New technology is not omnipotent. Keep a clear mind, and don't blindly pursue fashion. Without a solid professional foundation, they will not get suitable applications no matter how many new technologies. It is crucial to do a perfect intersection and integration between different majors and disciplines.

Modern information technology includes artificial intelligence. In addition to the "networking" feature, "intelligence" is another feature (although it is still in its infancy). Therefore, modern information technology can also be called intelligent technology broadly. So what exactly is artificial intelligence, and how does it relate to intelligent technology? These will be the content to be introduced below.

1.4 A Preliminary Understanding of AI

Artificial Intelligence (AI) is a fascinating topic. In a sense, its philosophical implications may be more substantial. Let us have a preliminary understanding of related issues related to "intelligence" and artificial intelligence.

(1) "AI+ everywhere"

"Intelligence" is a fashionable vocabulary, and it has become a hot spot of the whole society. There are numerous products such as "intelligence phones", "intelligence washing machines," etc., and concepts such as intelligent manufacturing, intelligence transportation, intelligent services, intelligent medical care, etc. These can be collectively referred to as "AI+", as shown in Fig. 1.11.

So the "intelligence" that all walks of life are talking about, whether there is substantive content, and where the "intelligence" of intelligent products is reflected, are questions that need to be calmly thought and answered. For example, various smartphones that are currently popular, according to merchants, their intelligence is mainly embodied in tracking user behavior, user authentication, emotion recognition, and natural language understanding. But if you analyze it carefully, it is not difficult to come to the conclusion that most of the current intelligence is more advanced automation, and it is still far away from the real intelligence requirements!

Fig. 1.11 Various "AI+"

Intelligentization is a bright path for the future development of many industries, but the concept of over-consumption is very dangerous. On the one hand, it will cause a lot of bubbles to be injected into the market, leading to unstable development; on the other hand, it will cause people to lose confidence in intelligence, and the development of all kinds of "AI + " will be fatally hit.

(2) The origin of the concept of machine intelligence-Turing test

The concept of artificial intelligence was formed at the Dartmouth Conference in 1956. It was called computer intelligence or machine intelligence. In AI, computers are also called machines. Generally speaking, the intelligence disciplines have achieved limited success in some aspects.

When it comes to the judgment of machine intelligence, a well-known "Turing test" can be used. The British logician and computer pioneer Alan Turing proposed a test in his paper "Computer and Intelligence" in 1950. He believed that instead of asking, "can a machine think," it is better to ask, "can a machine pass the intelligent behavior test." Suppose a machine can start a dialogue with a human but cannot be recognized whether a machine or a human. It is intelligent.

Figure 1.12 illustrates that the Turing test is divided into two stages. In the first stage, the tester distinguishes between men and women by asking various questions to the testees. The man's task is to make the tester think he is a lady, and the lady needs to prove to the tester that she is a lady. The computer replaces the man in the second stage, and the task remains unchanged. Suppose the computer has successfully deceived the tester the same number of times as the man has successfully deceived. In that case, the computer passes the intelligence test, and the machine is proven to have human intelligence.

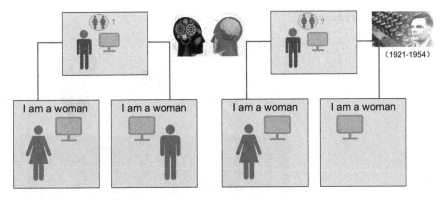

Fig. 1.12 The Turing test

Turing believed that 30% of machines should pass the test by 2000. Unfortunately, no computer has passed the test at this writing. It is very difficult for software developed using current artificial intelligence principles to pass the Turing test.

Although no machine has passed the Turing test yet, this does not hinder the exploration and research of artificial intelligence. Artificial intelligence is generally understood as: giving machines the ability to think and deal with problems like humans. The post-Turing AI enables machines to well-design tasks that even outperform humans.

(3) Human intelligence and artificial intelligence

For machines to have human-like intelligence, in the final analysis, we still need to figure out what human intelligence is, especially how human intelligence is generated, so that it is possible for machines to simulate.

To make machines have human intelligence, we still need to define intelligence. Human intelligence is multifaceted and includes language, mathematical, spatial perception, motion, music, interpersonal relationship, self-cognition, and natural cognition (Fig. 1.13).

The performance of human intelligence is diverse. Some people are good at mathematics and music, but some are good at music but not mathematics. There are very few talents who have a lot of knowledge. The current artificial intelligence is also similar.

Although the machines have not passed the Turing test, there are a lot of advancements in AI-related computing. For example, AI may outperform humans in repetitive work, complex calculations, and document retrieval. This is also true for intelligent construction, where machines can do many technical tasks automatically and efficiently.

(4) Does the machine think?

If you want a machine to have human intelligence, it must have the ability to think like humans. Therefore, the actual core issue of AI is whether the machine can think! It can be discussed from two aspects.

Fig. 1.13 AI versus human intelligence

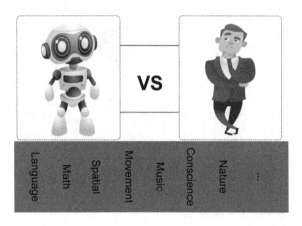

Firstly, one must answer why people think. If we can understand how thoughts are produced, we can create machines that mimic humans. Unfortunately, the current research on human beings is still minimal to provide a complete answer. Being able to think is a complicated problem for neural science and philosophy. This question may involve another generation of consciousness, which involves cognitive science, philosophy, and religion.

Secondly, one needs to know how the machine works. Current computers reply to "0" and "1" operations, and their combinations are driven by software code. Therefore, machine intelligence is the intelligence of software. However, the current software code cannot think on its own. In summary, machines currently cannot think like humans.

To sum up, the current machine cannot think like a human (but it is indeed much better than humans in some aspects, such as computing, etc.), and probably will not in the future unless the machine itself can generate true consciousness.

Figure 1.14 shows the two directions of AI development. In the foreseeable future, AI will still be dominated by "weak AI" instead of "Strong AI." Machines will fully surpass and control humans. It is just a scene in science fiction and movies. We don't need to worry too much!

A machine is a machine, after all. Although it has certain intelligence, its comprehensive IQ is far from that of humans, and even the IQ of some animals cannot keep up. The current AI only assists human beings but cannot completely replace human beings and does not possess real human intelligence. Humans need intelligent machines to perform complex tasks that humans are unwilling to do or difficult to do. But different times and people have different understandings of this "complex work".

Practically speaking, it may never be possible for a machine to reach the level of intelligence as a human being. Still, if it can assist human beings in performing some mental work, it may be enough. For intelligent construction, we are faced with engineering and technical problems.

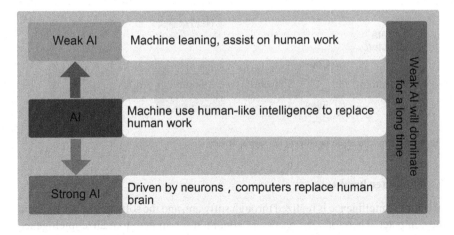

Fig. 1.14 Two development directions of artificial intelligence

All AI-powered products use the "weak AI" method. This approach works very well in data-rich domains. Whether it is speech recognition, face recognition, or simple machine translation, chess games, machines have done a good job. For example, in simple machine translation, the answer is relatively easy to get together, find enough data to match, and then adjust the order. However, it is still unsuitable for complex sentences as it involves the problem of understanding its meaning, and the same is true for human–computer dialogue. These require machines to really think, which they can't currently do.

(5) **How to implement intelligence**

Generally speaking, the realization of intelligence for an item requires computer hardware and software, as well as the participation of sensors. This is the indispensable "Three Musketeers", as shown in Fig. 1.15.

In the process of realizing intelligence, the hardware of the computer is the "brain", the sensor is the "organ" for acquiring information, and the software (control software and intelligent algorithm software) is the "soul". A brief description is given below.

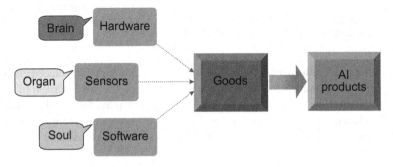

Fig. 1.15 Schematic diagram of AI products

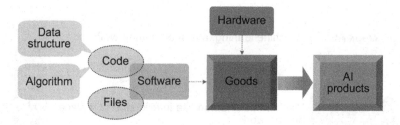

Fig. 1.16 Items are made intelligent through software

1. **Computer hardware**

All current intelligence is realized through software, and the software needs to run on the computer. Therefore, the necessary condition for an item to realize intelligence is to have the participation of computer hardware. As an intelligent "brain", the single-chip microcomputer is generally used at present, which is a kind of small microcomputer with a wide range of application fields.

2. **Sensors**

Sensors are controlled by software to perceive or obtain information. There are many types of sensors for specific requirements.

3. **Software**

Software is the "soul" of intelligence. Hardware is considered software's carrier, while software drives the hardware to work. AI functions are accomplished through software, which controls sensors and actuators simultaneously.

Figure 1.16 shows that software code written with data structure and algorithm is compiled into the software. Then, the software is embedded in the hardware to function.

Writing AI code requires data structures, intelligent algorithms, and specific applications. It often requires the cooperation of multiple disciplines. The popular programming languages for AI coding include C++, Java, Python, etc.

> Python is an object-oriented, interpreted programming language invented in 1991. Python's syntax is concise and can connect modules compiled in other languages. Python is widely used in data analysis and visualization, AI, etc., due to the availability of many open-source libraries.

Whether it is hardware or software, as long as the technology is advanced enough, the required performance and functions can be achieved. The advancement of hardware is to reduce the size of electronic equipment and increase the speed of calculation; the advancement of programming technology allows us to have more convenient programming tools. However, for AI, these are not the most critical. The real key is the algorithm!

Algorithms are processing steps that use computers to calculate or solve problems (somewhat similar to recipes). If the problem we face can establish a mathematical model, then it can be solved according to traditional programming methods. But if you are only dealing with a large amount of data, without a mathematical model, you can only rely on the computer to find the laws in the data by itself. This is machine learning. Therefore, algorithms are the key to achieving intelligence, but their essence is mathematics. This is where we need to be clear.

"The code is easy to write, but the model is difficult to build". For the problem to be solved, you first need to know what its model (physical model, mathematical model, and other models) is, and then you can start writing code. When a model cannot be established, one can try to use some kind of intelligent algorithm to establish a computer model based on a large amount of data available and give a solution.

All intelligent products now contain various forms of computers, the core of which is still the problem of machine intelligence. The level of intelligence is very limited, more like advanced automation.

For example, the basic function of a refrigerator is to cool according to the set temperature and keep it at the set temperature. If we want to get more functions-understanding the quantity of food in the refrigerator, reminding the shelf life of the food, reminding the proper diet, and receiving the food information of the refrigerator through mobile phone text messages, etc., it is necessary to implant a single-chip microcomputer, sensors and control the device, and then formulate the corresponding algorithm and write the corresponding program according to the demand.

But, if you analyze it carefully, the above-mentioned so-called smart refrigerators still lack real intelligence. Just have some automation functions, which is also the essence of most current smart products. A true smart refrigerator should learn to actively "think" to meet the needs of users, as do other products. But, when this demand can be realized is still unknown. Nevertheless, this new type of refrigerator is still much easier to use than traditional refrigerators. There is no need to entangle the degree of intelligence and to take a relaxed attitude towards the meaning of intelligence.

(6) Intelligent technology and "AI+"

Intelligent technology is a product of multiple disciplines. It can be understood as a broad term for modern information technology (Sect. 1.3), which mainly includes computer technology, perception technology (sensing and testing), automation technology, control technology, communication technology, and many technologies such as network technology and artificial intelligence (Fig. 1.17). Intelligent technology has relatively loose requirements for the degree of intelligence, and it does not care much about passing the Turing test.

The combination of intelligent technology and professional technology is "AI+", such as intelligent manufacturing, intelligent medical care, intelligent transportation, etc. This is an emerging, interdisciplinary, cross-domain comprehensive technology, which includes both common technologies (perception technology and machine learning, etc.) and individual technologies (professional technology, etc.), affecting

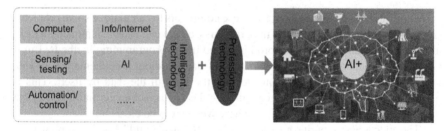

Fig. 1.17 Intelligent technology and "AI+"

almost all fields. Although this book introduces intelligent construction, its basic ideas and methods are also applicable to other fields.

1.5 Understand Intelligent Construction Technologies

Intelligent construction is the product of intelligent technology integration with the traditional construction industry. Although it is just emerging and its architecture is not perfect, it has attracted widespread attention. Let us have a general understanding of what intelligent construction is. From the current point of view, it is not important whether the definition is perfect, but understanding the connotation is the key.

(1) **Features that "AI+" should have**

The purpose of AI is to allow computers to do things instead of humans. Therefore, analyzing the process of human beings doing a project helps to understand the meaning of intelligent construction, which is also the basic problem faced by "AI+".

Before a person does a project, he must first understand the relevant information. This is the process of information "perception", then the information needs to be "analyzed" (induction, reasoning), and then "decisions" are made based on certain criteria, and finally "Execute". Do it yourself, ask someone else to do it, or not do it.

The reason why people have the above-mentioned abilities is that they have the ability to learn, to master relevant knowledge through learning, and to learn the ability to do projects. If machines were to replace people, machines should learn and perform "Perception," "Analysis," "Decision-making," and "Execution"—PADE, Fig. 1.18 illustrates the mirrored human learning and "machine learning" as shown in Fig. 1.18. Of course, the machine also needs to have the ability to learn, through learning, and master the ability to do projects. This is machine learning.

Can machines really replace humans in some projects? The answer is yes, but it requires the participation of other technologies. The computer is still the "brain" in the process. The four parts of PADE will be introduced below.

1. **Perception**

The purpose of perception is to obtain information. The computer itself does not have the ability to perceive information. It needs to use perceptive terminal equipment (sensor + data collector) to perceive, which is controlled by the computer.

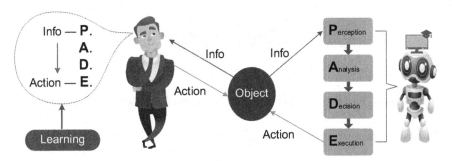

Fig. 1.18   Use machines instead of humans to perform tasks

2. Analysis

The information from the sensing terminal is in the form of data. The analysis is the process of extracting useful information from data, the purpose of which is to provide a basis for decision-making. Computers are good at analyzing data, but they need to be equipped with this ability first. This is the task of machine learning.

3. Decision

Decision-making is the process of deciding on an action plan based on the results of analysis. If you let a computer make decisions, you also need to let it master relevant knowledge and have the ability to make decisions. This is also the task of machine learning.

4. Execution

Execution is how to do the project, which is related to professional knowledge. At present, this part of the work has basically achieved mechanization and automation, such as mechanized construction, automated measurement, etc., and its operation is also controlled by a computer.

After a simple analysis, it can be found that the basic steps of replacing people with machines can be summarized as "Perception, Analysis, Decision-making, and Execution". These four steps are proposed based on summarizing what human beings do. They are universal to a certain extent and are the primary route for applying intelligent technology in various fields.

The "perception" part is in charge of the perception terminal, which mainly involves perception technology and has many commonalities. The "analysis and decision-making" is in charge of the computer, which is the main body of AI, and mainly involves various machine learning algorithms, which also have commonalities and are used in specific applications. It needs to be combined with domain knowledge. The "execution" part involves the most extensive knowledge. The content of different fields is different, which needs to be determined according to the specific situation, but the control technology has a commonality.

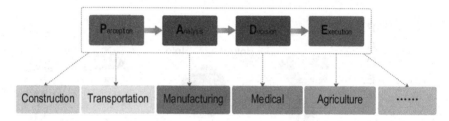

Fig. 1.19 The fundamental characteristic of AI+

A lot of practice has proved that no matter what field, the basic steps for people to do things are the same, but the specific content is different. The same is true for machines. Therefore, "perception, analysis, decision-making, and execution" are the basic characteristics of "AI+", which has specific applicability to all fields, as shown in Fig. 1.19.

The basic feature of "AI+" is sequential, that is, the process by which machines imitate humans doing things. The common technologies are the parts of "Perception, Analysis, and Decision-making". When applied in different fields, they need to be integrated with professional knowledge. For example, the integration with engineering construction is intelligent construction, which has already been applied initially.

(2) **The meaning of intelligent construction**

Regarding intelligent construction, although there is no unified definition (it is not a good thing to define new things prematurely), it is not difficult to get the basic meaning after understanding the basic characteristics of "AI+".

Through the integration of modern information technology and construction technology, autonomous perception, learning, reasoning, analysis, decision-making, execution, and control can be carried out in the entire process of construction (design, construction, and maintenance) or key links. It can adapt to environmental changes, optimize certain goals, improve the quality and efficiency of engineering construction, serve the entire life cycle, and finally realize an intelligent transportation infrastructure. Intelligent construction can be understood from the following aspects.

1. **Multidisciplinary integration**

The technology of the traditional construction industry is relatively mature, but it is precisely because of the mature technology that limits the further improvement of engineering quality. Modern technology has already caused a huge impact on traditional industries, and a comprehensive technology update is required.

"Stones from other mountains can be used for jade", the introduction of new technologies in the traditional construction industry and the development of modern construction technologies are inevitable trends. Judging from the current development situation, in addition to AI (which accounts for a small proportion), many technologies have been or are being integrated, such as computer technology, sensing

Fig. 1.20 Related technologies for intelligent construction

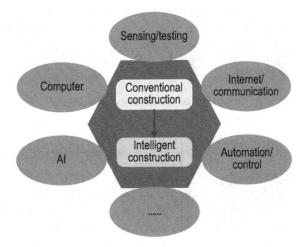

and testing technology (perception technology), automation and control technology, communication and network technology, etc., as shown in Fig. 1.20.

In addition to the above technologies, emerging technologies such as the Internet of Things, big data, and cloud computing will all be useful. In fact, with the development of science and technology, more and newer technologies will be incorporated into engineering construction.

Judging from the current situation, in addition to the few applications of AI in engineering construction, several other technologies have begun to be applied. Some have become popular, such as computer technology, sensing, and testing technology. However, intelligent construction should be reflected in "intelligence", and how to combine existing construction technology with AI will be the focus of research.

2. **Should have the basic characteristics of "PADE"**

Intelligent construction should have a certain degree of autonomous perception, learning, reasoning, analysis, decision-making, communication, coordination, execution, and control capabilities, which can be summed up as "perception, analysis, decision-making, and execution". This book (including other sub-volumes of the series) will also follow this technical route. In fact, the essence of intelligent construction lies in this, which is also the fundamental difference from "automation" technology.

The difference and connection between intelligence and automation. Automation originally refers to the use of mechanical actions instead of manual operations to automatically complete specific tasks. With the development of science and technology, especially the widespread application of computers, the concept of automation has been expanded. The use of machinery not only replaces human physical labor but also replaces or assists part of mental labor. Intelligence refers to the intelligent completion of specific tasks on the basis of automation. At present, many so-called intelligences are actually still more advanced automation, but the two tend to be assimilated.

Fig. 1.21 Examples of automatic survey technologies

Fig. 1.22 An automated construction site that uses total station technologies

In engineering construction, it is one of the goals we pursue to replace some human labor (physical labor and mental labor) with various intelligent machines. If machinery replaces human physical labor, it belongs to automation, and only human brain labor is replaced by intelligence.

Figure 1.21 shows some applications of total stations, laser 3D scanners, and unmanned aerial vehicles (UAVs). Compared with traditional measurement methods, it saves a lot of manpower and material resources. There are already some shadows of intelligence, but it is not really intelligent. It is a work in progress.

Traditional road construction mostly uses manual operations. With the advancement of science and technology, mechanized construction has become the main method. Figure 1.22 is a schematic diagram of mechanized road construction. Some of the mechanical operations are carried out under the guidance of instruments, which are already automated operations, but they are not in the true sense of intelligent construction.

The characteristic of intelligent construction is the use of computers to replace part of human mental work. Therefore, AI is one of the key technologies in intelligent construction, and it is necessary to understand and master AI knowledge.

3. **The long-term goal is to improve project quality and efficiency**

At present, many industries are advocating the concept of full life cycle, and there have been concepts such as full life cycle roads and full life cycle railways. Specific to the implementation level, the first thing is to improve the quality and efficiency of project construction and serve the entire life cycle. This is both the primary and long-term goal of developing intelligent construction. Building a smart and intelligent transportation infrastructure will be the ultimate goal. Of course, this process will be very long and requires the joint efforts of multiple disciplines. The current development of intelligent construction is to improve the quality and efficiency of the project. Otherwise, it is just empty talk.

Intelligent construction is still a new concept. This concept evolved from intelligent compaction during the construction phase. Rolling compaction is the main method of filling construction, and quality control is the key. Intelligent compaction evolved from continuous compaction control. Based on the signals continuously sensed during the rolling process, the performance parameters of the compacts can be obtained through complex analysis, which can realize "perception, analysis, decision-making, and execution" process. Extending the concept of intelligent compaction is intelligent construction, extending to the stage of survey, design, and maintenance, which promotes the concept of intelligent construction.

In 2016, Prof. Guanghui Xu of China, Dr. George K. Chang and Prof. Nazarian of the United States, and Prof. Antonio G. Correia of Portugal founded the International Society for Intelligent Construction (ISIC). The goal is to create a platform of intelligent construction for the world to advance and implement ICT. Please find out more about ISIC on its website (www.IS-IC.org). Frankly speaking, there was no in-depth thinking about the meaning of intelligent construction at that time, and the original intention was not to limit international organizations to the field of intelligent compaction. However, with the deepening of research, a new understanding of the concept of intelligent construction has been gained, and it has gradually been felt that intelligent construction is the only way to improve the quality and efficiency of the project in the future.

Intelligent construction is still in its infancy. From informatization to networking, automation and intelligence, it is changing the traditional construction methods. Although there is still a long way to go from real intelligent construction, it is already on the way and worthwhile. We would pay attention and participate in it.

1.6   The Keys to Implement Intelligent Construction Technologies

In the early stage of the development of intelligent construction, there are many things that need to be explored, especially the rapid development of modern technology, and many new and fashionable technologies have appeared, but not all technologies can be applied immediately. There is a gradual process that needs to be grasped. To resolve the main contradictions, concentrate on making breakthroughs at some "points."

(1)   Engineering data is key

The "intelligence" in intelligent construction is an application of AI, which requires autonomous perception, learning, analysis, decision-making, and action. The core is to use machines to complete these tasks, but machines need to use a large amount of data for training (learning) before they can work. The acquisition and accumulation of engineering data become the key.

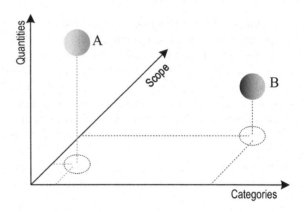

Fig. 1.23 The meaning of engineering big data

Big data (also called massive data) is a very fashionable vocabulary and one of the indispensable foundations of intelligent technology. Its magnitude is calculated as "PB (1 PB = 1024 TB, 1 TB = 1024 GB)". But for engineering construction, the "big" of engineering big data is not simply a large number, but more importantly, there are more data types and a wide range.

In Fig. 1.23, the number of data set A is much larger than that of data set B, but its type and range are smaller than that of data set B. In terms of engineering, the quality of data set A is not as good as that of data set B, including there is not as much useful information as B. The criterion for determining the quality of data is not the absolute quantity but whether it contains enough useful information.

For example, in the quality control of roadbed compaction, the traditional method is to detect sampling points and use samples to infer the population. If continuous testing is used, quality data (that is, the overall) can be obtained for all points, which is a kind of big data compared to sampling testing. Although the number is much smaller than the "PB" level, it is a population rather than a sample. For the engineering field, data that contains all the facts of an event is big data (it may be more appropriate to call it full data).

Engineering data is one of the keys to determining the quality of machine learning. Without sufficient data support, it will not be possible to complete the training of the machine. In layman's terms, engineering data can be compared to "experts with rich engineering experience (knowledge)", and its role in engineering construction is self-evident.

Engineering data not only supports the training of machines but is also the source of information about the entire life cycle of roads and railways. Whether it is a road or a railway, it enters the operation period after the survey, design, and construction phases. As time goes by, the quality of the "road" will continue to decline, and after reaching a certain level, it will enter the maintenance and repair period. This is actually the beginning of the secondary design and construction, but the scale is different. If intelligence can be realized in the whole process of construction, then a large amount of useful data (linear, material, structure, performance, function,

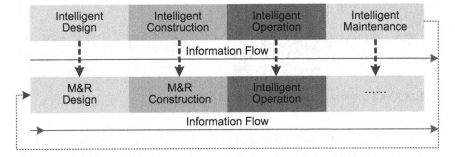

Fig. 1.24 Information flow in intelligent construction

environment, etc.) in the various stages of design, construction, operation, and maintenance will form an information flow (Fig. 1.24), you can extract a lot of knowledge from these data to promote the improvement of design, construction, and maintenance level, which in turn will promote the development of intelligent construction, which is also of great benefit to intelligent transportation during the operation period.

How to obtain engineering data is also a problem that everyone is more concerned about, mainly related to perception technology. Of course, the experience and knowledge of various engineering and technical experts is also a key link that cannot be ignored. How to collect it will be described later. For a more detailed discussion, please refer to other parts of the series.

In addition, corresponding to big data is the problem of small data, which should be paid more attention to. In many cases, it may not be possible to obtain enough data. How to train the machine based on small data (small samples) will be more challenging.

(2) **Innovate with science and technology**

Intelligent construction is not to set aside traditional construction technology and develop another set. It needs to be developed on the basis of traditional technology and then integrated with intelligent technology. Otherwise, it will become a "tower in the sky." Nevertheless, we should be soberly aware that traditional construction technology has reached a bottleneck period, and it is difficult to advance the construction industry only relying on traditional professional knowledge.

The traditional construction industry is based on classical mechanics (the same is true of modern science and technology). However, the development of modern science and technology is based on modern and even modern physics. Traditional construction technology needs to add these modern basic theories as a new driving force for development. Otherwise, even if some new technologies are introduced, it will be difficult to achieve greater development. There is no way to talk about intelligent highways or railways.

(3) **Construction phase is key**

Intelligent construction covers the whole process of engineering construction. The ideal approach is to achieve intelligence in the three stages-design, construction

Fig. 1.25 Construction site focusing on paving and compaction

and maintenance, and to drive the intelligentization of related equipment such as construction machinery and testing equipment. However, judging from the current development situation, it is unrealistic to realize intelligentization in the whole process of construction, and a process is needed. A more realistic approach is to make breakthroughs in key links first and then expand to other links.

Inspecting the whole process of project construction, it can be found that if the intelligence is realized from the design stage, the actual significance is not great. Because even if the intelligent design is done well, without the cooperation of construction, it is impossible to form a good "product". If you start with intelligence from the maintenance and repair stage, it is impossible to achieve the purpose of improving the quality of engineering construction. The construction phase is the main battlefield of engineering construction, the manufacturing process of products such as "Road" (Fig. 1.25), and the place where construction technology is concentratedly displayed. Therefore, choosing the construction process as a breakthrough to realize intelligence is not only feasible, but also necessary.

The construction process of road and railway engineering is mainly based on the rolling of various fillers. The key to intelligent construction is to first realize the intelligentization of the rolling process. Judging from the current development situation, the "road construction technology" represented by intelligent compaction has initially realized intelligence. As the forerunner of intelligent construction, intelligent compaction completely includes the characteristics of "perception, analysis, decision-making, and execution". Therefore, starting from intelligent compaction, from point (intelligent compaction) and surface (intelligent construction), and then extended to the design and maintenance stage, we can gradually realize the intelligentization of the whole construction process.

1.7 Framework of Intelligent Construction Technology

Highways and railways are linear structures for vehicles (cars and trains), which are obviously different from other structures. In order to build the basic framework of intelligent construction technology, it is necessary to understand the whole process of construction and see from a global perspective what content can be prioritized for intelligentization.

(1) **Understand the roadway construction process**

Whether it is a road or a railway, the construction process is basically the same, which can be collectively referred to as "road" construction. A complete construction process is divided into three phases-design phase, construction phase and maintenance phase. In addition, there are pre-planning stages, etc. The planning stage is not considered here. For planning issues, technologies such as intelligence and virtual reality must be very useful.

Let's take Fig. 1.26 as an example to understand the construction process of the highway. If a road is built from A to B, the whole construction process and main work content are clear. As bridges, tunnels, and traffic engineering ancillary facilities are responsible for related majors. I won't repeat them here.

1. **The elements of the design stage**

The central task of the design stage is to design construction drawings. Generally speaking, after a road is determined to be built, its key control points (necessary locations) are determined. Therefore, the topography and geomorphology related to the direction of the line must be measured first. Then the preliminary line selection will be carried out to survey the geological conditions along the line. After determining the route direction, a series of work such as linear design, cross design, material design, structural design and drainage design are carried out, as shown in Fig. 1.27.

The topography is measured by various measuring equipment, and data can be obtained to draw and encrypt topographic maps. This is the first work to be carried out in the design stage. Speaking in fashionable language, this part of the work is called perception, and these devices are perception terminals (this formulation is to prepare for the Internet of Things involved later).

Figure 1.28 shows some working pictures of surveying and surveying. It can be felt that these fieldworks are still very hard. How to realize the mechanization, automation, unmanned methods and intelligence of field surveying and surveying may be an urgent problem to be solved at present.

Fig. 1.26 Build a new infrastructure road from A to B

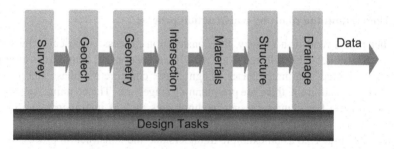

Fig. 1.27 The elements of the design phase

Fig. 1.28 Examples of survey devices and technologies

After the survey task is completed, the next task is to "talk on paper" in the room. The current design work has changed from manual drawing in the past to computer-aided design (CAD), and some 3D technologies have also begun to be used.

The linear design includes plane design, longitudinal section design, and cross-section design; cross design (intersection with other roads and railways) mainly includes plane crossing and three-dimensional crossing design; material design includes design of cushion, base, and surface materials (strictly, in other words, subgrade materials should also be included); structural design includes two parts: subgrade and pavement. They are a whole and should not be designed independently. There have been many cases of diseases that have occurred due to negligence of subgrade structure; drainage design includes Plant road surface water, road structure internal water and groundwater drainage design; in addition to these designs, road environment (landscape) design also needs to be considered; correspondingly, there is also the preparation of project budgets.

The above-mentioned designs are all based on CAD. The computer plays a decisive role. All data relating to the design will be stored in the computer and flow to the construction stage (Fig. 1.27) to form a data flow (information flow) for the construction stage transfer.

With the development of technology, emerging technologies such as 3D visualization and virtual reality will be developed. Building Information Modeling (BIM) is one of them, but for filling structures, it has not been well applied and developed. It is very necessary to develop unique three-dimensional design technology for highways and railways. It is not only necessary to carry out three-dimensional and dynamic simulation of geometric shapes, but also to carry out the dynamic simulation, calculation, analysis, and simulation of structural performance, and also to combine with intelligent technology.

2. **The elements of the construction phase**

As a "product", like industrial products, it is also necessary to turn the design drawings into "products" in accordance with design requirements. This process is construction (manufacturing). Construction is the main battlefield of engineering construction. The main work content of road construction is shown in Fig. 1.29.

Construction preparation involves a wide range of content, including training the construction team, checking design documents, restoring alignment, retesting the route, and preparing construction organization. Coordination and overall planning should be carried out on the information platform from the perspective of systems engineering.

Construction staking is to place the geometric shape of the structure on the drawing on the actual site as the basis for construction. An automated stakeout has been realized. Before the construction of the base layer and the surface layer, it is also necessary to carry out construction staking.

For the road cutting, it needs to be excavated to form the roadbed structure. For the embankment, it is necessary to first perform the necessary treatment of its foundation part, and then fill and roll it in layers according to the selected filler and construction machinery. Compaction quality control is the key link. Intelligent compaction technology has been used to solve this problem.

The base construction is somewhat different from the roadbed construction, and the material control is stricter. The materials are different, and the construction process is not exactly the same. But basically it can be divided into material production, transportation, paving, rolling, shaping, and health preservation.

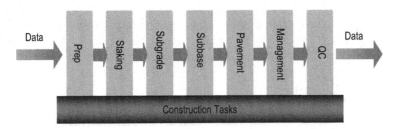

Fig. 1.29 The elements of the construction phase

Surface construction is divided into two types: asphalt surface construction and concrete surface construction. Asphalt surface course (hot mix type) construction mainly includes four links: mixture production, transportation, paving, and rolling. Concrete surface construction includes concrete mixing, transportation, paving, vibrating or rolling, curing, etc., as well as treatment of joints.

Construction management is a complex systematic project involving personnel, materials, machinery, quality, schedule, safety, finance, etc., throughout the entire construction process. Using a systematic approach (see the next chapter for details) for overall management is a good choice, and intelligent technology will play an important role.

Quality control is an indispensable and important link in the construction process, involving many aspects. Among them, compaction quality control is one of the keys. Intelligent compaction control technology has been used to solve this problem.

After the completion of the construction, various technical data will be formed, which will be transmitted to the maintenance management department in the form of data flow (information flow) (Fig. 1.29). This is a batch of very valuable engineering data, which is the initial state set of the road during the operation period. If some sensors can be embedded in the structure, long-term monitoring of the structure's performance can be realized, and it is also very good training data for machine learning.

3. **The elements of the maintenance stage**

After the construction is completed, it enters the operation period. The main problem is intelligent transportation, which is beyond the scope of this book. From the perspective of construction, operation and maintenance coexist, restricting and affecting each other. The main task of the maintenance phase is to determine the maintenance scale, cost, and construction time based on relevant data. The main tasks include road condition detection and prediction, comprehensive analysis, overall optimization, maintenance decision-making, maintenance design, and maintenance construction, etc., as shown in Fig. 1.30.

The road conditions mentioned above, including the geometric shape of the structure, internal defects, and performance degradation, need to be detected by certain methods, and the sensing (detection) technology will be very useful here. Condition prediction is based on the judgment of the current status and the future status, which is

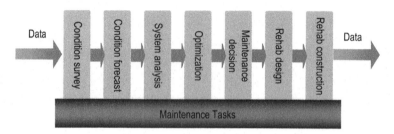

Fig. 1.30 The elements of the maintenance stage

the main basis for formulating medium and long-term maintenance plans; comprehensive analysis is for various data. In addition to performance parameters, other aspects of data should also be considered, such as historical data (passed during the construction period), etc.; overall optimization is an optimization based on comprehensive analysis and taking into account multiple factors. Generally, "cost-effect" is the optimization goal; the maintenance plan is made based on the optimization results. Decide to determine the repair scale and time. The next step is to enter the second design and maintenance, but the scale is different.

For railways, the roadbed part is basically the same as the roadbed of the highway, except that the implemented standards and specific technical requirements are different. The upper track structure is obviously different from the pavement structure, which also leads to different contents of the later maintenance work, but the basic process of construction is the same.

(2) Intelligentization of engineering construction

After briefly reviewing the construction process of "Road", let us now see what technologies can be intelligent. In principle, all technologies can be intelligent, but there will be differences in the degree of intelligence. However, in the early stages of development, we should still grasp the key points, not the beard and eyebrows.

1. Design stage

The purpose of the design is to design construction drawings that meet technical standards. Intelligent design is to allow the machine to design a design drawing that meets the requirements (currently in digital form stored in a computer) based on the perceived data and in accordance with technical standards, like a human engineer. A complete intelligent design should include the four basic processes of "perception, analysis, decision-making, and execution", as shown in Fig. 1.31.

Perception: For design, the topography, geology, traffic volume, climate, economy, and other data along the line are indispensable. Surveying is conducted to collect and obtain the above data by terrain surveying, geological surveys, traffic volume, or economic surveys. All the above can be considered perception technology.

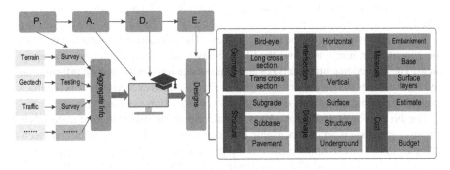

Fig. 1.31 Intelligent content and basic process in the design phase

Analysis: The computers can independently analyze the perceived data and generate several design drafts according to the requirements of technical standards.

Decision: Decision-making is the result, while analysis is the process. The computer analyzes and compares the design drafts, makes decisions, and obtains the optimal design plan.

Execution: The execution of the design phase should be effortless. The computer would generate the design file based on the decision. These design data will then "flow" or are transmitted to the construction phase.

Before a machine performs intelligent design, it should first have the ability to learn like a human. A computer needs to autonomously learn design-related knowledge (machine learning), master design skills, and reach the level of independent design.

It needs to be explained that the perception technology itself also has intelligence problems, and it can also be carried out in accordance with the process of "perception, analysis, decision-making, and execution". Such as remote sensing, after the image is sensed, a computer can be used to perform intelligent analysis and decision-making to identify the relevant information in the image.

2. Construction stage

The task of the construction phase is to build structures (roads and railways) in accordance with the design requirements. It is the product manufacturing phase, and the construction site is the "production workshop". Intelligence should revolve around how to "build roads" intelligently. For the railway, the roadbed construction is the same as that of the highway, and the track part is a prefabricated mechanized construction (automatic track laying machine), which has been automated. The following is an example of road construction.

For road engineering, whether it is a roadbed or a pavement, it is a problem of rolling loose bodies into structures. Therefore, the key technology is to select suitable fillers and fully roll them in accordance with the design requirements. Regarding fillers, it involves particle size, shape, gradation and moisture content, etc. How to achieve intelligent control is still a topic that needs to be studied. For rolling, the current intelligent compaction technology has begun to play a role. The intelligentization of the construction phase still follows the four basic steps of "perception, analysis, decision-making, and execution" (see Chap. 7 for details), as shown in Fig. 1.32.

Figure 1.32 is mainly for the rolling process. The interaction information between the construction machinery and the filling body is sensed by the measuring equipment. The trained machine analyzes and makes decisions on the sensed information, and finally issues the execution instructions. The so-called execution here is feedback control, including the control of packing and construction machinery.

For the base layer and surface layer, the mixing and transportation of materials in the mixing station is an indispensable link, and there are also many intelligent technologies in it. For example, in the mixing process of materials, various sensors can be used for sensing (gradation, temperature, etc.). Then the computer will comprehensively analyze these data, and then make decisions and issue control instructions to the mixing equipment.

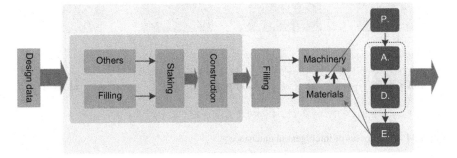

Fig. 1.32 Intelligent content and process in the construction phase

As for drainage engineering, support engineering, and traffic engineering, if it is to be intelligent, it is necessary to realize mechanized construction first and then configure sensing equipment and computers, etc. On this basis, intelligent construction can be carried out. In addition, after the construction phase is over, the relevant data will flow to the operation and maintenance phase, which is very important engineering data.

3. **Maintenance stage**

The main task of the maintenance phase is to determine when to carry out maintenance and the scale of maintenance based on relevant data. Therefore, the data is the most critical at this stage, and its source is shown in Fig. 1.33.

The maintenance data mainly comes from three aspects-the data transmitted at the end of the construction, the data obtained by the sensors embedded in the structure, and the data obtained by various on-site inspections. These data are an important basis when formulating a conservation plan.

The intelligence of the maintenance stage is mainly based on formulating maintenance plans and still follows the four basic steps of "perception, analysis, decision-making, and execution", as shown in Fig. 1.34.

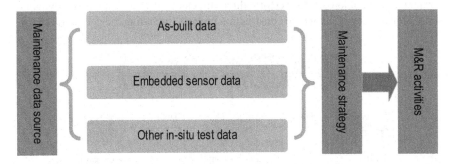

Fig. 1.33 Source of maintenance and rehabilitation data

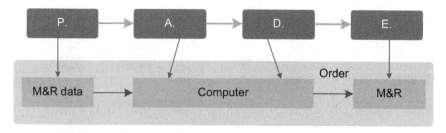

Fig. 1.34 The process of intelligent maintenance

Data plays a decisive role in the formulation of intelligent maintenance plans. Accumulating maintenance data is very important for the evaluation and management of road projects, and we need to pay enough attention to it.

(3) **Intelligent construction technology framework**

Many technologies in the construction of transportation infrastructure need to be intelligently upgraded. However, in the early stage of intelligent construction, it is still necessary to make key breakthroughs and refine some key technologies to form the main structure of intelligent construction technology, and then gradually add various technologies. Improve the technical system.

According to the previous analysis, the intelligent optimization in the design phase, the intelligent compaction in the construction phase, and the intelligent decision in the maintenance phase can be selected as the main skeleton to form the basic structure, as shown in Fig. 1.35.

The selection of intelligent optimization, intelligent compaction, and intelligent decision-making as the main framework of intelligent construction technology has covered the key technologies in the whole construction process. These technologies include intelligent perception (such as measurement, and detection), intelligent control (such as construction machinery), intelligent identification (such as filler, and distresses images), etc., which will be further explained later.

In addition, many emerging technologies have also been applied. For example, virtual reality technology can perform three-dimensional visualization and dynamic simulation display of the whole construction process (including the geometric shape,

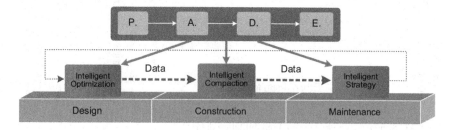

Fig. 1.35 Intelligent construction technology framework

structure, performance, function, construction process and operation of buildings, etc.), and network technology can helping to realize the connection and cooperation between objects, these are worthy of our attention. But these are all applied technologies, and what really works is the knowledge related to the profession, especially the basic theory.

Chapter 2
Systematic Approach to Representing Objects

Abstract In this chapter, the essential characteristics of the systematic approach are given from the perspectives of definition, structure, integrity, feedback, function, performance, behavior, stability, and system model. On this basis, the fundamental problems of the system are given. The technical characteristics of the black box analysis method and the method of mastering the new technology are introduced. Then, the systematic method is adopted to establish the road structure system, artificial intelligence system, and intelligent construction technology system to demonstrate the P.A.D.E. principles. The ability to execute and the chapter's fundamental principles of building and information platform system are explained at the end.

2.1 Definition of the System

System theory is an emerging cross-cutting science. The most basic ideas are integrity, interconnection, and evolutionary development. There are many definitions of a system, but the consistent meaning is that a system is an integrated network with a specific function(s) achieved by different constituent elements (components) related to each other and creating a relationship with the environment.

(1) **Systematic expression**

Generally, the system components that cannot be further subdivided are called the elements or elements of the system. The elements and their connections constitute the system. It can be expressed as system $=$ < feature set, association method set > , or symbols.

$$S =< A, R >　\qquad (2.1)$$

where, A-represents the set of elements that build the system, and the number of elements should not be less than two; R-represents the association method between elements.

© China Railway Publishing House Co., Ltd. 2023　　　　　　　　　35
G. Xu and D. Wang, *Introduction to Intelligent Construction Technology of Transportation Infrastructure*, Springer Tracts in Civil Engineering, https://doi.org/10.1007/978-3-031-13433-3_2

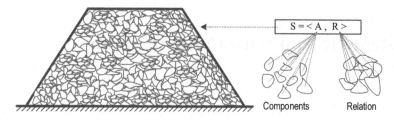

Fig. 2.1 Structure system composition of the filler body

Using set theory to express the system with a specific algebraic structure can lay the foundation for quantitative analysis. The Formula (2.1) shows that the system is jointly determined by component elements and association methods between elements. The system elements and their interrelationships determine the essential characteristics of the system.

For example, Fig. 2.1 shows a filler structure system (such as a roadbed structure), mainly composed of filler materials and the interconnection between filler particles. Here A = {particles of various sizes and specifications} ∪ {water}; R = {occlusal} ∪ {interlocking} ∪ {friction} ∪ {bonding}.

If the quantity of system elements or the number of system element types is large, the system can also be divided into several parts for research. These parts are called subsystems, and they all have unquestionable integrity, as shown in Fig. 2.2. Dividing the system into several subsystems is more conducive to in-depth research in many cases.

For example, the road structure system S can be divided into several sections according to the length of the road. Each section is a subsystem (Si). It can also be divided into three subsystems: surface layer, base layer, and roadbed according to the structure type, as shown in Fig. 2.3.

(2) **System structure**

The system structure refers to how the various elements within the system are related. The R in Formula (2.1) is the mathematical expression of the structure.

Fig. 2.2 Systems and subsystems

$$S = S_1 + S_2 + S_3 + S_4 + S_5$$

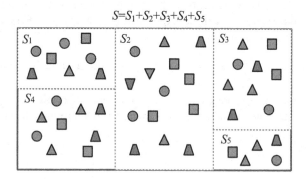

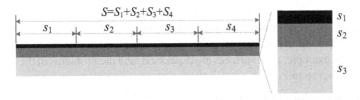

Fig. 2.3 Road structural systems and subsystems

It is not the number of elements that determines the complexity of the system structure but the types and associations of the elements. For example, Fig. 2.1 shows that the system element is the filler particles, divided into several categories. Still, there are only four association methods: occlusion, interlocking, friction, and adhesion. The particles are organically combined by these association methods to form a whole.

Generally speaking, the quality of the system structure is reflected by its performance indicators. For example, the road structure system's modulus is one indicator that reflects the structural capacity.

(3) **Integrity of the system**

System elements are related to each other and form a network together. Its feature is integrity, and it emphasizes the overall effect. Therefore, integrity refers to only the system's characteristics but not the elements or isolated parts. Integrity can be expressed as "the whole is not equal to the sum of parts."

Integrity is the core viewpoint and principle of system theory. Once the elements constitute a system, it has the nature of integrity that the element does not have, and the integrity provides more functions than the sum of the parts ("$1 + 1 > 2$" principle). Figure 2.4 illustrates the difference between system integrity and parts (the sum of elements). If there are no interactions between the stones, it will not have the overall structure and properties; if there is no connection between the battery and the light bulb, it will not light up.

The general function of the system is not equal to the simple sum of the function of the elements but is determined by the interaction between the elements. This way of thinking is different from the traditional concept of reductionism: dividing the system into many simple parts, inspecting them separately, and then adding them together.

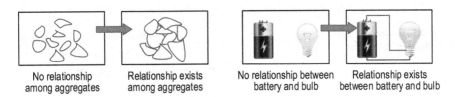

No relationship   Relationship exists   No relationship between   Relationship exists
among aggregates   among aggregates   battery and bulb   between battery and bulb

Fig. 2.4 Integrity of the system

Instead, the system should be treated as a whole, revealing the overall nature of the system from the relationship between the elements.

System thinking is essential to analyzing the structural system of highways and railways. For example, road surfaces are distressed, and the problems in the subgrade may cause uneven rail track.

(4) System environment and feedback

The external factors connected with the system are called the system's environment. The so-called connection here refers to the exchange of matter and energy between the system and the environment. The environmental effect on the system is called the input (or stimulus), and the system effect on the environment is called the output (or response).

Another critical concept is feedback, which is the reaction of system output to the input. Specifically, it can be divided into negative feedback and positive feedback. Negative feedback means that the information fed back is in the opposite direction of the input information and inhibits the original development direction of the system. Positive feedback means that the information fed back is in the same direction as the input information and strengthens the original development direction of the system.

Input, output, and feedback are fundamental concepts in control theory (see other parts of the series for details). Figure 2.5 shows the relationship between them.

Generally speaking, any system is produced in a particular environment and then runs, continues, and evolves in a particular environment. The system's characteristics depend on the environment, and its properties will be altered with environmental changes. The road structure system follows this typical system principle.

(5) System function and performance

The role of the system is called the function of the system. System performance refers to the characteristics and capabilities after the interrelation between the internal elements of the system, which are inherent to the system itself. Performance decides how the function will be achieved.

Take road engineering as an example. The primary function of the road structure system is to provide a stable and safe operating environment for the superstructure and driving, such as the surface function of the road surface. The physical and mechanical properties of the road structure define its multiple performance categories, such as strength, rigidity, stability, etc. The performance of the road structure plays a decisive role and provides the function.

Fig. 2.5 Environment and feedback of the system

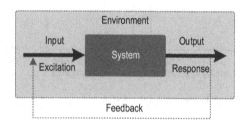

(6) **System behavior**

The changes in the system relative to the environment are called the system's behavior. It is a system change related to the environment that reflects the degree of external influence on the system.

Different systems have different behaviors, and the same system will have different behaviors under different conditions. Behavior performance mainly includes learning, adaptive, evolutionary, and dynamic behavior. Understanding the system's behavior is vital to capturing the system's mechanism. For example, any change in the behavior of a road structure system is closely related to its environment (loading, water, temperature) and its performance.

(7) **System stability**

System stability refers to the system's ability to maintain its original state under particular interference. For structural engineering systems, structural stability is the most critical issue.

No matter what type of system it is, it is crucial to maintain sufficient stability. Once an unstable system deviates from its normal state (the system's state can be observable and recognizable), it can no longer be restored to its original state. The deviation will increase, and the original system will lose some characteristics.

According to the control theory, the system stability is due to a series of harmful feedback mechanisms or self-regulation mechanisms within the system. However, this adjustment mechanism has certain limits. Once the limit is exceeded, the system will collapse.

The stability of the road structure system shown in Fig. 2.6 refers to the ability of the physical and mechanical properties of the structure without significant changes over time under the action of vehicle loads and natural factors, such as water infiltration and high-temperature environment. Suppose the external effect of the road structure system exceeds the ability of self-adjusting (exceeding the threshold specified by the structural rigidity and strength). In that case, the original stability will be destroyed, and the system's overall function will undergo significant changes, further reducing the structural performance and function. The stability during the construction period is low, and the compaction aims to improve its stability. At the initial stage of operation, the system's stability and state are at best; after a period, the system's

Fig. 2.6 Changes in stability of roads' structural system

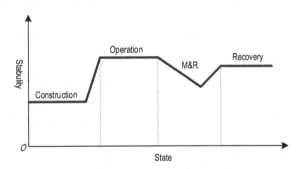

stability will gradually decrease. After maintenance and repair are performed, its stability will be restored.

(8) System evolution

The evolution of a system refers to the changes in its properties (structure, performance, function, stability, behavior, and state) over time. Any system will change over time.

The power of system evolution comes from the joint action of internal and environment. Internal factors refer to how system elements are related; environmental factors refer to changes in the system's inputs. The results of evolution are divided into two types. Evolution is changing system attributes from low-level to high-level and straightforward to complex. The system will be more stable; on the contrary, degradation is the opposite, and the system may collapse. In the system evolution process, generally, it will not be smooth sailing, and there will be degeneration in the evolution, which will develop in waves. The development history of AI systems is a good example, as shown in Fig. 2.7.

(9) System model

The so-called model refers to a simulation product obtained by depicting the system's main characteristics by adopting an appropriate form of expression. Various models can describe the system's characteristics. The mathematical model is most important, whereas the computer model has only been developed recently.

1. Mathematical model

The mathematical model describes the interaction between the internal elements of the system, the system itself, and the environment. A mathematical model is a tool for quantitative analysis of the system. In addition to the commonly used equation forms, such as geometric figures, algebraic structures, topological structures, and other forms can also be implemented. For example, the transportation network and

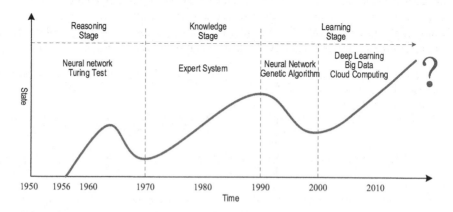

Fig. 2.7 Evolution process of artificial intelligence system

social network models use graph theory. The system's relationship between inputs and outputs is widely described as a quantitative analysis model.

In engineering systems, the commonly used mathematical models are various equations. Algebraic equations mainly model static systems, such as truss structures. Differential and difference equations are mainly used in continuous and dynamic systems. For example, pavement models use partial differential equations. Whether the differential equations can be used is worth studying.

How to build a model is very important for a specific engineering problem. It is necessary to simplify the model based on the primary focus without covering every detail. Achieving precise simplification requires the accumulation of essential knowledge in the knowledge domain.

Not all problems can be converted into mathematical problems, and there are many problems that mathematical formulas cannot express. For this type of problem, perhaps a computer model is a solution.

2. Computer models

The use of computer modeling is a hot topic in the systematic approach. The model defined by a computer program is usually called a computer model. For example, expert systems, machine learning (such as neural networks), etc., are all computer models.

All mathematical models can be transformed into computer models by writing programs. Systems that cannot build mathematical models can also build models through computers. Various machine learning algorithms are computer models based on data.

In addition, some complex systems that actual experiments cannot test can also be tested by computer simulation, called digital experiment technology, a computer model. Virtual reality, cyber-physical systems, digital twins, etc., have emerged in recent years and are also computer models. Using computer simulation, a small number of experiments can achieve ideal results, which many industries have successfully used. Such models should be paid special attention to intelligent construction.

2.2 Basic System Problems

The specific characteristics of different systems are different. However, as a system, its fundamental problems have commonalities. This has a guiding significance for us to master a particular subject's knowledge or scientific research. It can enable us to clarify the clues quickly, cut into the subject, and clarify the essence of the problems we are facing. The systems mentioned here mainly refer to artificial systems, such as transportation infrastructure, artificial intelligence, and intelligent construction systems.

If the system is regarded as a product, there will be fundamental components of product design, manufacturing (processing and production), sales, use, and service (maintenance). Inspired by this, the system's fundamental problems can be deduced by analogy.

Figure 2.8 shows that if two of the three factors (system, stimulus, and response) are known, the rest can be obtained. This leads to three fundamental system analysis, identification, and control questions. These three fundamental issues are the main tasks determined by the system, incentives, and response. They are also the main issues in control theory. Based on the understanding level of the system structure, it can lead to the "white box, black box, and gray box" issues, which will be discussed separately later.

For manual systems, system design and manufacturing issues must be added. Therefore, the artificial system has five fundamental problems: design, manufacturing, analysis, identification, and control, covering three stages of design, manufacturing, and service, as shown in Table 2.1. With more detailed consideration, system maintenance issues can also be added. Of course, the maintenance phase can also be regarded as another manufacturing phase. In addition, there are system management issues and so on.

For a specific system, the fundamental issues are not equally important. For example, the system analysis problem should be the most important for the existing

Fig. 2.8 Relationship between system, input, and output

Table 2.1 Fundamental problems and examples of the system

Problem	General systems		Road structure system		Stages
	Known	Solution	Known	Solutions	
System design	Excitation, response	New system	Loading, displacement	Road structure Design	Design
System construction	System elements	New system	Various materials	Road structure	Construction
System analysis	Excitation, system	Excitation	Loading, road structure	Stress, strain, displacement	In-service
System identification	Excitation, response	System	Load, displacement	Modulus, density	In-service
System control	System, response	Excitation	Road structure, response	Load control (over-load issue)	In-service

road structure system. The following takes the road structure system as an example for further explanation.

(1) Design issues

In the case of known traffic loads, by rationally selecting road structure (including materials) parameters, designing and optimizing a structure that meets the response (displacement) requirements is the main task of the design stage.

(2) Manufacturing issues (construction issues)

The road structure system manufactures the "product" of the road structure based on the design data. The focus of road construction is paving and compaction.

(3) Analyze the problem

Suppose the parameters of the road system (modulus, density, Poisson's ratio, structural layer thickness, etc.) and vehicle load are known. In that case, the stress, strain, and displacement responses can be solved according to the mechanics theory to understand the behavior of the road structure.

(4) Identify the problem

The so-called parameter identification is how to solve the attribute parameters of the road structure (such as modulus, density, etc.) under known input and output conditions. This work is an inverse problem, which is challenging. Compaction quality control in the construction process belongs to this type of problem, and various non-destructive inspections also belong to this category.

(5) Control issues

The essence of the control problem is to control its input type and size based on the known characteristics of the road structure and the allowable displacement response. The problem is not prominent on the railway due to the relatively fixed vehicle load. In road transportation, controlling overloaded vehicles fall into this category.

Through the above simple analysis, we can see that the systematic method is beneficial for us to control the overall situation and clarify the attributes of the problem. Cultivating a systematic way of thinking will also help the research and apply intelligent construction technology.

2.3 The Black Box Method

The systematic approach is a methodology that can help us analyze problems from different angles and perspectives and has universal significance. In actual work, there is a so-called "black box" method, which is helpful for us to deal with many problems and master new technologies. Such a method is worthy of attention. The following examples illustrate this method.

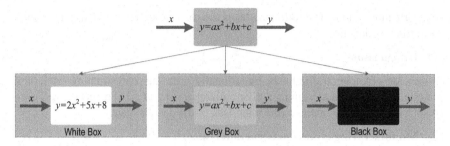

Fig. 2.9 Three analytical methods

Suppose the structure of a system is a quadratic function ($y = ax^2 + bx + c$), the input is x, and the output is y. Our understanding of this system explains the black box method (see Fig. 2.9).

If we know the system very well, for example, we know its structure (quadratic function) and parameters (a, b, c), we call this system a "white box." This type of system is the easiest to master and is what we most hope for. According to the input, the output can be directly obtained. For example: $y = 2x^2 + 5x + 8$, if $x = 5$, then $y = 83$.

If we only partially understand the system, for example, we only know that the structure is a quadratic function and the coefficients are unknown, we call this system a "gray box." For the gray box problem, the first step is to determine the coefficients of the system (parameter estimation in system identification) and then solve the output. Regarding parameter estimation, the most commonly used method is the least square method, which solves the coefficients based on input and output data. In addition, the AI method can also be used. It is unnecessary to solve the coefficients, and the AI model will directly give the results. The gray box problem is common in engineering. The use of FWD to back-calculate the modulus of road structure is a typical example.

If we don't know anything about the system, for example, we don't know the structure and parameters of the system, then we call this system a "black box." The black box problem is also universal. Many unknown problems can be attributed to black boxes, such as human brains, universe planets, and sealed instruments. AI also belongs to this type of problem, and the internal operating mechanism of neural networks is a typical black box problem.

Facing the black-box problem, there is a universally applicable analysis method- the black box system identification method. This method is to identify the properties of the black-box functions, characteristics, structure, and mechanism by observing the correspondence between the input and output of the black box without opening the black box.

A typical example in life is choosing watermelons, as shown in Fig. 2.10. Watermelon can be regarded as a black box system, tapping is the input, and the output is the sound from the watermelon. An experienced person recognizes the maturity of the watermelon based on the tapping and sounds. This is called black-box recognition.

Fig. 2.10   Black box method to identify watermelon ripeness

An in-depth analysis of this example will reveal a black box identification problem, a learning problem and big data problem, and a dynamic excitation problem, which contains a lot of scientific and technological knowledge.

We know that not everyone can recognize maturity by tapping a watermelon. Only experienced people can. This "experience" is obtained through many tapping practices (learning). "Tapping practice" is sample training for the selection of watermelons, and "experience" inevitably requires a large amount of tapping practice (big data) as support. Machine learning is a method that will be described in detail later.

The black box method identifies the system based on input and output data. This method is closely dependent on data. The more data, the richer the types. The wider the scope, the better the results will be. The current AI is the same.

There is also a similar simulation method for reference. If a system can be designed under the same input, its output is the same as or similar to the black box's output. Then the system realized the simulation of the black box. A successful example is using an oscillator circuit in electricity to simulate the vibration equation in mechanics.

In 1956, William Rose Ashby of the United Kingdom explained the black box method more systematically in "Introduction to Cybernetics." He believes that "the black box problem appears in electrical engineering. Give the engineer a sealed box with input and output connectors on it. Apply a certain voltage to the box, then observe, which can infer the situation inside the sealed box."

In addition to the uses mentioned above, the black box method has multiple other uses. If we open our minds, we can apply this method to learn and master new technologies, especially for modern information technology in intelligent construction, because intelligent construction contains too many new technologies and involves multiple majors. This question is related to setting up professional courses that will not be discussed in detail here. So how do you quickly master new technologies? The following examples can provide some insights.

For general users, the TV is a black box. Generally, they don't know anything about the internal structure, but it does not affect the use; mobile phones and computers are also black boxes and do not affect the use. Although these devices' internal structure and performance are a black box problem, they can be used as long as the users know their functions and operations. Inspired by this, it is also feasible to use the black box method to learn and master new technologies, which can quickly achieve the purpose of application.

Fig. 2.11 Learn new technology using the black box approach

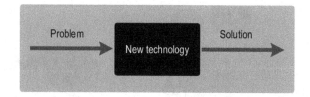

There are two ways to learn new technology. The first is to be familiar with the relevant knowledge of the technology in deep; the second is to use the black box method, which only cares about the input of the question and the output of the answer. New technologies in intelligent construction are divided into two categories. The first one is hardware technology, such as various electronics and control equipment; the second one is software technology. Whether hardware or software, we can all regard it as a system. It may be a white box for those proficient in electronics and software. They know the system's internal structure and can design it, but it is a black box problem for most civil engineering professionals. The core of using the black box method to learn and master new technologies is to focus on the performance and functions of the new technology, the black box system, without investigating its internal structure. It is enough to know how to input questions, what answers to output, and whether they are reliable, as shown in Fig. 2.11.

The black box method exists in all career fields, but not everyone has understood it from a systematic perspective. Various testing instruments, CAD drawing software, etc., can be mastered by the black-box method for general users. This method is often used when judging the quality of testing equipment.

Modern testing instruments are all electronic equipment, which can be regarded as a black box, by inputting information (such as sine wave) into the instrument and then judging the quality of the instrument according to the output information (whether it is a sine wave). Of course, if we can master these electronic technologies, we can open them and check them ourselves. The black box may become a white box or become a gray-box problem.

2.4 Road Structure System

The systematic approach looks at and solves problems from a global perspective. Road engineering is a field that we are more familiar with. Let's take the construction of a road structure system (the railway engineering structure system can be analogized) to analyze how to use a systematic approach to solve traditional problems. Figure 2.12 shows the composition of the road structure system.

The elements (A) of the road structure system (S_R) are the roadbed, the base layer, and the surface layer. The association mode (R) is the connection between the roadbed and the base layer and the connection between the base layer and the surface layer, which is the structure of the road. S_R can be expressed in the following form.

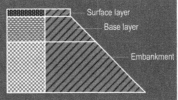

Fig. 2.12 Composition of road structure system

$S_R = <$ (subgrade, base layer, surface layer), (subgrade and base layer connection, base layer and surface layer connection) $>$.

The input of S_R is vehicle load and environment factors (moisture and temperature); the output is stress and displacement. The system, input, and output are clearly defined. Table 2.1 makes it easy to determine its fundamental problems.

Figure 2.13 shows the five fundamental problems of the road structure system, including system design, system manufacturing, system analysis, system identification, and system control.

Take system design as an example. The geometric road design is the main task of route selection, including graphic design, longitudinal section design, cross-section design, etc. The road structure design is to determine the types and appropriate methods of the three subsystems of subgrade, base, and surface. The road's material design determines the component elements and the way of association. Specifically, it can be divided into the design of the three subsystem components: the roadbed, the base layer, and the surface layer. The division of road structure system design is shown in Fig. 2.14.

All issues should be apparent when using a systematic approach to re-examine the road design. This approach will help us grasp and analyze the issues from a global perspective. At the same time, it is also perfect for realizing intelligent design and guiding us on where to start introducing intelligent technology. Other problems of the road structure system can be deduced by analogy.

It should be noted that the system method is also applicable to subsystems and system elements, but they need to be regarded as another system. For example, the

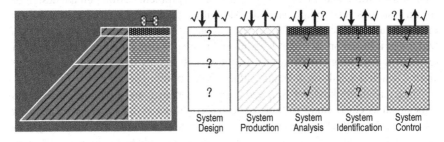

Fig. 2.13 Five problems of road's structural system

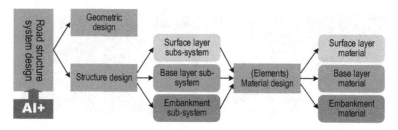

Fig. 2.14 Division of road structural system design

roadbed subsystem has five fundamental problems: design (construction), analysis, identification, and control [5]. For material design, a systematic approach can still be used. For example, the asphalt mixture design can be regarded as a material design system, and there will be fundamental problems.

The system can be built for railway engineering following the method mentioned above. For example, a ballastless track system's elements are roadbed, gravel track beds, sleepers, and steel rails; the elements of a ballastless track system are roadbed, concrete base slabs, track slabs, and steel rails. Interested readers can construct it by themselves.

2.5 Artificial Intelligence System

The current AI is a system of software and hardware (S_{AI}). It consists of two parts: a computer and a program. Their correlation is mainly reflected in the machine (hardware) and the program (software) and between the programs. It can be expressed as follows:

$S_{AI} = <$ (hardware, software), (machine and program, program and program) $>$

The main problem of S_{AI} is how to carry out system design and system manufacturing, including selecting algorithms, programming, testing, and learning, which will be discussed in detail later.

The core of S_{AI} is to autonomously make judgments or predictions based on given input information (actually, it is a system analysis problem). Although the design of such a system is very complicated, it is straightforward to use, as long as you master the essential characteristics of S_{AI}. For example, in face recognition (Fig. 2.15), S_{AI} will judge it based on the input face information and output the correct recognition result. But designing such a system is not easy. It mainly depends on the learning ability, which depends on data, algorithm, and computing power.

There are many kinds of S_{AI}, and their performance and functions are not the same, such as chess game systems, speech recognition systems, image recognition systems, etc. An intelligent construction system is an artificial intelligence system proposed for engineering construction, integrating multiple technologies and having specific performance and functions. How to build such a system will be challenging.

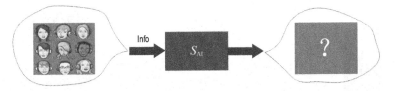

Fig. 2.15 Face recognition

In addition, S_{AI} can be used according to the black box method for general users. The more advanced S_{AI}, the easier it will be to use.

2.6 Intelligent Construction Technology System

Many factors are involved in constructing transportation infrastructures, such as personnel, technology, equipment, materials, information, capital, construction mode, etc., and "people, money, and material" will be the foundation of construction. From a system perspective, the complete intelligent construction system (S_{IC}) is:

$S_{IC} =$ <(personnel, technology, equipment, funds,…), (interconnections between elements)>

This is a vast and complex nonlinear system. The input is the land, and the output is transportation infrastructure such as roads and railways, as shown in Fig. 2.16.

The construction and operation of intelligent construction systems such as management are generally government actions beyond this book's scope. From a technical point of view, the following is only to build an intelligent construction technology system.

(1) **System composition**

Intelligent construction technology includes traditional construction technology, computer technology, sensing and testing technology, automation and control technology, network technology, artificial intelligence, etc. These technologies constitute the elements of the technical system, and they are all related to each other (this is the so-called interdisciplinary), forming an organic whole, as shown in Fig. 2.17.

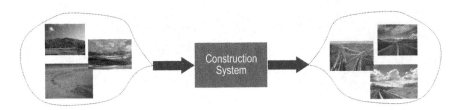

Fig. 2.16 Transport infrastructure based on the input "land"

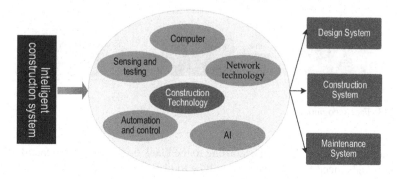

Fig. 2.17 Intelligent construction technology system composition

The intelligent construction system is vast and complex, and the content involved involves multiple disciplines. Here, the boundaries between disciplines are broken. Building such a system is not an easy task. We need to understand and master more knowledge. It is not enough to rely on specific professional knowledge. We must learn the ability to use multi-disciplinary knowledge comprehensively. The systematic approach gives us enlightenment. Regarding new knowledge as a system, you can apply it as long as you know its performance and functions and know-how to input and output.

The intelligent construction system can be divided into intelligent design, intelligent construction, and intelligent maintenance systems. Computer technology plays a leading role in testing, control, calculation, drawing, etc. All use computers as tools and are the basis of intelligence. Figure 2.18 shows the three critical subsystems.

Since the construction stage is the main field of engineering construction and the product manufacturing process, filling and rolling during the construction process is crucial, which is also an important reason why intelligent compaction is generally concerned by the engineering community. With the intelligent compaction subsystem as the center and information flow as the medium, the basic structure of the intelligent construction system is formed by expanding forward to design and backward to maintenance, as shown in Fig. 2.18.

The information flow (data flow) in Fig. 2.18 has different meanings at different stages. The information transmitted from the design to the construction stage is mainly construction drawings. The information transmitted from the construction and maintenance stages is mainly the structure's performance. The information in

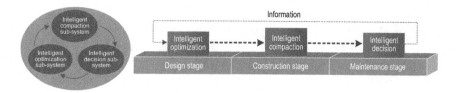

Fig. 2.18 Major subsystems of an intelligent construction technology system

Fig. 2.19 Perception of external information using instruments

the maintenance stage is mainly the comprehensive data of all aspects of the structure, which will be transmitted to the rehabilitation design stage.

(2) Basic functions of the system

Building an intelligent construction system aims to integrate many technologies into a whole, hoping that it can partially replace or assist humans in doing something. Therefore, the intelligent construction system's function is a question we do more concerned about. The preliminary analysis has been carried out in Sect. 1.5, and some explanations will be made here.

1. Perceptual ability

Perception is the key to obtaining data and information. Whether a person or a machine, information is obtained through a particular way of perception. Humans perceive external information through visual, auditory, tactile, and other means; the primary means of mechanical perception is through perceptual devices (instruments), as shown in Fig. 2.19.

It is closely related to sensing technology, from surveys to various detections in engineering construction. For intelligent construction, perception technology is the primary means to obtain information related to sensors and data collectors and is closely related to professional knowledge. In many cases, various models required for perceiving information must be established.

2. Analysis ability

Analysis and decision-making capabilities are the core of the intelligent construction system and the primary embodiment of intelligence. The task of analysis is to intelligently analyze data, extract useful information, and provide a basis for decision-making.

How can the system have analysis capabilities? This involves machine learning. The intelligent construction system can only make correct analysis and decision-making on the information after learning (training) and obtaining enough "knowledge." In many cases, learning and analysis coexist. The so-called "learning by doing, doing learning by doing" is the best portrayal. Learning, analysis, and decision-making capabilities are unique functions of all intelligent systems.

3. Decision-making ability

Decision-making ability refers to the computer's skills and ability to participate in decision-making activities and alternatives. It is also one of the main characteristics

Fig. 2.20 Split of the
intelligent construction
system by function

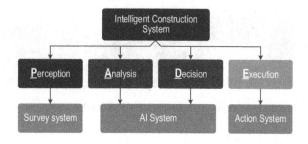

of intelligent construction systems. The premise of decision-making is to analyze and extract useful information; the basis of decision-making generally comes from domain knowledge, relevant regulations, and some prior knowledge.

4. **Execution ability**

After the system has undergone perception, analysis, and decision-making, it has reached the stage of taking action and executing it. There are many titles for intelligent construction systems, such as measurement, design, drawing, construction, calculation, detection, and control. The level of execution ability reflects the level of doing things.

The intelligent construction system can be divided into several subsystems according to different stages and functions, as shown in Fig. 2.20.

The perception part can be assembled into a measurement subsystem. The essential elements include sensors and data acquisition systems, further divided into intelligent and general types. The analysis and decision-making parts can be combined into the aforementioned AI system. The execution part is an active system depending on specific issues.

The intelligent construction system is complex. For users, there is no need to master how to build such a system in detail, as long as you grasp the system's characteristics and the know-how to input and output results are available. You must master knowledge in multiple systems development fields and cooperate with the team to achieve expected results.

2.7 Information Platform System

The systematic approach emphasizes connection and integrity. Intelligent construction is a complex and huge system that runs through the entire process and all links of project construction. Its subsystems are distributed in a decentralized manner. A dispatch center must integrate these subsystems for unified application and management. The current feasible method links various parts to form an interconnected whole through network technology. The information platform system or the intelligent construction system is the prototype of the engineering Internet-of-Things. Cyber-physical systems are also developed as the foundation of the digital twin.

Fig. 2.21 Various networks

(1) **Network and information platform**

Networks refers to communication and computer technology connecting various electronic terminals distributed in different locations. They communicate according to a specific network protocol to share software, hardware, and information and establish mutual exchanges of various information such as text, images, audio, etc. The existence of networks can be felt in transportation, finance, business management, education, business, and family life, as shown in Fig. 2.21.

The essence of the Internet is to connect and share. With the development of network technology, the Internet of Everything has become a trend. The network and interconnection of the intelligent construction system can be realized through the information platform. The scattered information can be unified on the platform. Then, the data-sharing mechanism can be established. On this basis, the comprehensive utilization of data can form "$(1 + 1) > 2$", the overall effect of the system.

All the electronic terminals mentioned here should contain a microcomputer so that it can be connected to the network. For the intelligent construction system, various technologies are closely integrated with computers. They have the technical characteristics of networking and the Internet of Things, which have laid the foundation for integrating them through network technology.

Information platform is a trendy vocabulary, and its essence is the existence of digital and networked information. In the information age, the understanding of information is usually based on the meaning of the information platform. Digitalization corresponds to computerization, and networking corresponds to the Internet. Therefore, in layman's terms, an information platform combines a computer and a network as a system that combines hardware and software.

> Digitization is a term widely used by the media (digital economy, digital city, and digital construction site.). The digitization of information is the computerization of information. Since the computer can only recognize 0 and 1, all information must be converted into binary code when entering the computer. This conversion process is a digitization process. In testing, analog-to-digital converters (A/D) convert analog signals into digital signals to facilitate computer recognition.

There are many information platforms, among which cloud platforms (Fig. 2.22) are surging, dominating, and are the mainstream of information platforms. "Cloud" (also a general term) is a metaphor for the Internet, and its concept is derived from

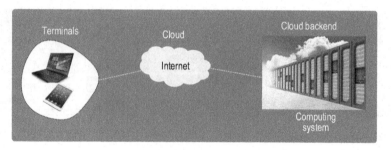

Fig. 2.22 Implications of the cloud platform

"cloud computing." The cloud platform can realize the interconnection and inter-communication between various terminal devices. All the resources users enjoy are provided by a cloud background (a computer system that is a cloud platform) with solid storage and computing capabilities.

Figure 2.22 graphically illustrates the composition of the information platform. The central computer system (host) has mighty computing power and storage capacity, integrating the required applications. The platform is also composed of programs that drive the connection between the host and various terminals (mini-computers, tablet computers, etc.). The terminal can apply to the host for various cloud services (online services), such as computing and querying.

The information platform is the bridge between the various subsystems and users for the intelligent construction system. All terminals (including subsystems and users) exchange information with the platform and transmit various data to the information platform, such as design information, construction information (materials, progress, quality, etc.), various testing data, etc. (Fig. 2.23).

(2) **Principles of constructing an information platform**

An information platform is also a system with some standard features from a system perspective. This is the basic principle that all information platforms must abide by.

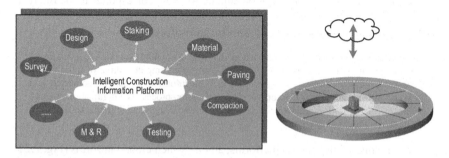

Fig. 2.23 Intelligent construction information platform architecture

1. **Hierarchical**

The information platform comprises three levels: basic information unit, intermediate information unit, and advanced information unit, with a bottom-up "box-like" structure. This structure facilitates the integration, splitting, and management of the platform.

2. **Interactivity**

The interactivity of the platform includes horizontal interactivity and vertical interactivity. Horizontal interactivity refers to the interconnection between different platforms; vertical interactivity refers to the interconnection between platforms at different levels.

3. **Unity**

The platform should have a unified information format, data interface, text protocol, operating environment, etc. The use and development of the information platform need to be carried out by unified standards to ensure horizontal and vertical compatibility and sustainable development.

4. **Openness**

Openness means that the platform's functions, structure, and levels can be changed. With the improvement of platform construction, its functions and information processing capabilities will be expanded and improved.

With the development of technology and the increase in demand, intelligent information platforms will be the critical component of an intelligent construction system.

Chapter 3
Empowering Machines to Learn

Abstract This chapter firstly introduces the basic idea of computer simulation of the human brain and analyzes the essential characteristics, implementation steps, and machine learning methods. It then focuses on the ANN method's learning process, introduces the weighting adjustment process in detail, and deduces the calculation method. A simple example is given. This chapter also includes a brief introduction to deep learning. On this basis, it introduces some other machine learning algorithms, including supervised learning, unsupervised learning, reinforced learning, etc. It finally analyzes the current artificial intelligence and illustrates the basic principles for applying machine learning in engineering construction.

3.1 A Computer Simulates Human Brains

Making machines have the ability to think and deal with problems like humans has always been the goal of AI. Although we are far from the goal, many researchers continue exploring new AI technologies. One of the most crucial questions is how to use a computer to simulate the human brain.

Since the concept of AI was proposed, there have been many ideas to simulate the human brain by computer. These ideas can be classified into two main types: (1) knowledge-based simulation as an expert, called symbolism, and (2) neural network simulation to create a similar physical structure of the human brain, called connectionism. In addition, there is a third way of simulating the human brain by intensive learning called behaviorism. We will discuss the first two types of methods in the following sections.

(1) **Knowledge-based expert system method**

The basic idea of an expert system is to allow computers to acquire more knowledge to simulate the human brain and perform like an expert. The expert system is a computer program whose development process is knowledge engineering. An expert system contains a large amount of the knowledge of many experts in a specific field, which can analyze, infer, and judge to simulate human experts' decision-making process

© China Railway Publishing House Co., Ltd. 2023
G. Xu and D. Wang, *Introduction to Intelligent Construction Technology of Transportation Infrastructure*, Springer Tracts in Civil Engineering, https://doi.org/10.1007/978-3-031-13433-3_3

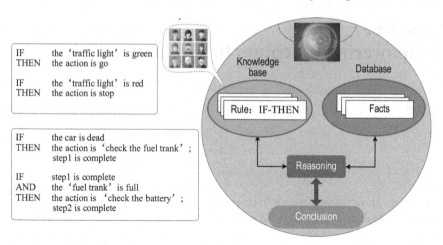

IF   the 'traffic light' is green
THEN   the action is go

IF   the 'traffic light' is red
THEN   the action is stop

IF   the car is dead
THEN   the action is 'check the fuel trank' ;
         step1 is complete

IF   step1 is complete
AND   the 'fuel trank' is full
THEN   the action is 'check the battery' ;
         step2 is complete

Fig. 3.1 Basic structure of the expert system

and thus solve complex problems. The expert system is a system with collective intelligence.

> Knowledge is the extraction of a human's understanding of the world. An expert is a person who has comprehensive, authoritative knowledge and experiences in a particular area, for example, roadway experts. Experts typically express their knowledge in the form of "rules."

In expert systems, knowledge is usually expressed in the form of "rules." "if–then" statements express rules. "IF" is the condition, and "THEN" is the conclusion. Some logical operations such as AND, OR, and NOT can be inserted in the middle of "IF" and "THEN." For example, "IF the traffic light is green, THEN you can go through" is a traffic rule, as shown in Fig. 3.1.

The rule-based expert system is the most commonly used method, whose basic structure is shown in Fig. 3.1. The most critical part is the knowledge base, which contains many rules (knowledge). Each rule expresses a relationship, advice, directive, policy, etc. When the condition (the IF part) of a rule is met, the behavior following the THEN is performed.

Inference machines (software, not hardware), another critical component, are the processors of knowledge connecting facts (information) to knowledge in the knowledge base. The inference machine matches the user's information (fact) input with the condition part of some rules in the knowledge base through the human–machine interface. After reasoning, the system makes decisions to solve a problem.

A database, also known as a fact library, contains information relevant to the problem. The inference machine matches the information in the database with a certain IF part in the knowledge base. If the IF part is satisfied, then the THEN part (behavior) is executed. For instance, if the information in the database is "the traffic light is red," the inference machine will search the corresponding rule in the knowledge base and conclude with a "stop" rule, as shown in Fig. 3.1.

The quality and quantity of knowledge in the knowledge base determine the level of an expert system. Knowledge acquisition is the key to the knowledge base's success and a typical bottleneck of the expert system design. How knowledge is acquired and becomes a rule requires careful consideration in a design process.

Expert system is the most critical and active application field in the early stage of AI, which is recognized as the breakthrough for AI from theoretical research to practical application and from general reasoning to applying specific knowledge to solve real problems. Despite its disadvantages, the expert system is still the most practical and widely used application of AI.

The storage of knowledge and applying knowledge to solve problems are two essential functions of an expert system. The expert system needs to specify the logical relationship between rules in detail humans, which thus makes the programming very tedious. In addition, because the logical relationship between rules is not apparent, the search will be prolonged when too many rules exist. The expert system is not intelligent since there was no autonomous self-learning ability. Until the 1980s, researchers realized that expert systems only applied to specialized areas and had no universal applicability, forcing them to develop techniques to simulate the human brain.

(2) Neural network method based on physiological brain structure

The human brain is made up of a large number of interconnected neurons. Another way to simulate the human brain is to simulate the physiological structure of the brain, which is called an Artificial Neural Network (ANN). Although ANN was proposed as early as 1943, it didn't develop until the 1980s.

The basic idea of the ANN model is that the brain can be regarded as a network composed of many interconnected neurons, and "wisdom" comes from the connections between many neurons. The same information can induce different actions with different connection strengths, which is also the main idea of "connectionism," as shown in Fig. 3.2.

It's important to note that this basic idea is just one of many ANN's ideas, and there's still little knowledge about how the brain works. However, the ANN model designed according to the basic idea mentioned above can be applied in practice, and its success is evident to all.

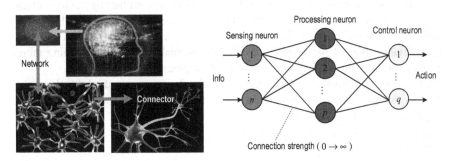

Fig. 3.2 Brain~neuron~activity

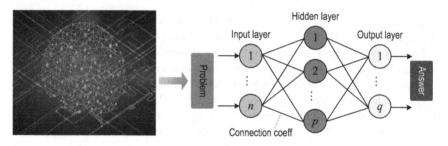

Fig. 3.3 The basic structure of ANN

The ANN model can only be treated as a mathematical model (a complex function with multiple layers) without considering the physical structure of the human brain, which inspired the development of the ANN model.

A typical ANN structure, as shown in Fig. 3.3, is a redesigned network structure based on the physiological structure of the human brain. Generally, it is divided into three layers: input layer, hidden layer, and output layer, composed of different numbers of neurons. The input layer receives inputs from the outside, passes them to the hidden layer for processing, then passes them to the output layer for processing, and finally outputs the results.

The strength of connections between neurons is different, like different connection coefficients (or weights). Because of the adjustment of these connection coefficients, the network as a whole element can make the correct output to match the input. This is how the ANN model works and is also the embodiment of the learning ability.

> Each node of ANN was treated as a neuron in the human brain simulation, and the network composed of these "neurons" was called a neural network. We can treat ANN as a black box (system) if we take a systematic approach. As long as we tell input and output, ANN can automatically learn the function relationship between input and output based on a large amount of data (but we don't know the specific expression). One of the theoretical foundations of ANN is that the three-layered ANN model can look for any function. As long as the amount of data and the number of neurons are large enough, ANN can learn automatically and provide the desired output.

In addition, a Genetic Algorithm (GA) is a kind of bionic simulation method that simulates the genetic, evolutionary mechanism of humans or other lives. GA is often used for optimization problems. Both ANN and GA are relatively mature algorithms and widely used. Especially, ANN's learning ability has made the machine succeed in simulating part of the human brain functions, making machine learning more popular and driving intelligent technology development.

It's important to note that the current AI technologies cannot think like humans. The reason may be that how the thinking works is not evident, and how consciousness is generated is still unknown.

3.2 How Do Machines Learn

Machine learning is one of the cores of AI. Machine learning allows computers to simulate human learning activities, replacing the human brain with learning and action abilities. Machine learning is a general term for several algorithms, such as artificial neural networks (ANN) and support vector machines (SVM). Some traditional statistical learning methods are AI technologies, such as cluster analysis, random deep forest, etc.

It is important to note that machine learning does not mean that the "machine" actually learns, executing the codes humans have written.

> AI has many new terms, such as "machine learning" and "deep learning." We should not be confused by these terms. We should look beyond the phenomena to see the essence. If you look closely at these terms, you will find that most of them, like "hardware, "is software. "Machine learning" refers to a type of algorithm, as does "Classifier," and "Support Vector Machine" (SVM) is also not hardware but an algorithm.

With the development of science and technology, hardware may have the learning ability in the future. It has been reported that researchers have been trying to develop a human brain model with a million chips. Although it is unknown whether it will succeed, this new idea of using hardware to simulate the human brain is being explored, which deserves attention.

(1) Human learning versus machine learning

Learning is an essential intelligent behavior of human beings. Although there is no unified definition, it does not limit the use of this vocabulary. A deep understanding of human learning methods and processes will significantly promote the development of AI. The following compares human and machine learning to illustrate machine learning more clearly.

Human beings accumulate a lot of experience through observation, and then after induction, they can acquire skills. When humans encounter unknown problems, they will use the acquired skills to make judgments or speculations. For instance, if you can recognize a dog, it is the experience accumulated through many observations. Therefore, from the beginning of observation to acquiring skills for human beings is "learning," as shown in Fig. 3.4.

For a machine, the learning is through the input data to accumulate knowledge. If the machine can use the knowledge to accomplish a given goal, it obtains a skill.

Fig. 3.4 Human learning versus machine learning

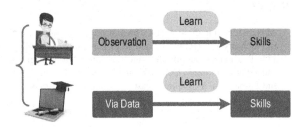

Thus, the process by which computers acquire skills from a large amount of data is called "machine learning."

Learning from experience, this concept was firstly proposed by Bayes in 1783. Bayes' theorem using historical data on similar events to determine the possibility of the occurrence of an event is a mathematical method of learning from experience. It is the basic idea of machine learning. In 1950, the Turing Test was the prelude to AI. After the concept of AI was proposed in 1956, machine learning was formally proposed in 1959. In 1997, IBM's Deep Blue beat the world champion in chess, drawing much more attention to intelligent technology. Alpha Go, developed by Google, focused on Go and eventually beat the then-Go champion in 2016, bringing AI to a new peak (actually, it is machine learning, but the outside loves the word "AI").

The core idea of machine learning is not complicated. Simulating the human learning process allows the computer to automatically obtain the ability to judge and predict with the accumulation of data (experience). However, it is very professional due to some excellent machine learning performance. It has become more and more popular after the hype of the media. More people are more or less interested in machine learning and misunderstand machine learning, and they think machine learning is mysterious and omnipotent. They blindly exaggerate the function of machine learning and ignore that machine learning is not omnipotent and has its scope of application.

(2) **Procedure of machine learning**

AI desires machines to do things like human beings, such as analysis, judgment, reasoning, prediction, decision making, etc. From a professional perspective, machine learning includes classification and regression tasks.

Classification is classifying information according to a particular attribute, such as the distinction between cat and tiger, face recognition, etc. Regression (also called forecasting) searches for hidden patterns based on available information to make predictions about new information, such as stock forecasts or weather forecasts. The information here is in a broad sense, generally referring to experience, things, data, behavior, etc., which are all "data" in the eyes of the computer. Figure 3.5 shows a schematic diagram of classification and regression.

If a mathematical model can be used to solve a problem, then this problem can be solved using traditional programming methods. However, suppose there is only data without a mathematical model. In that case, the machine has to find the regularity in the data by itself, and this process is called machine learning (training). When new data is input into the trained model, it will automatically classify or regress it (Fig. 3.5).

AI is divided into three stages. The first stage is computational intelligence – fast computing and memory storage; the second stage is perceptual intelligence – visual, auditory, tactile, and other perceptual abilities; and the third stage is cognitive intelligence – the ability to understand and think. Cognitive intelligence is the most significant gap between machines and humans in AI, making reasoning and decision-making by matching learning hard to come true.

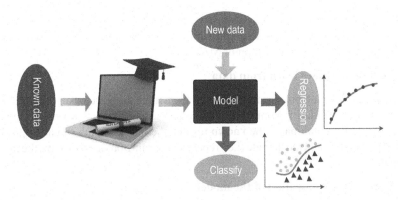

Fig. 3.5 Machine learning implementing classification and regression

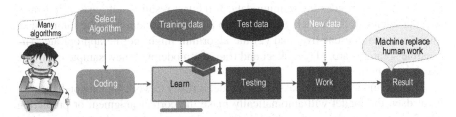

Fig. 3.6 Steps of machine learning

The essence of machine learning is that programmers write programs to learn data, which must be clearly understood to avoid misleading by those fashionable terminologies. For a specific problem with a batch of data, the general steps for machine learning are shown in Fig. 3.6.

1. **Preparing training data**

In machine learning, training data is needed to compare the training results with the known results. Training data is known as sample data, also known as label data, which provides the correct answer. The quantity and quality of training data directly affect the quality of the model.

Data acquisition is a very complex problem, requiring perceptual technology, and the cost of acquiring training data is often staggering.

2. **Choosing the correct algorithm**

Many machine learning algorithms exist, such as ANN, SVM, etc. Each algorithm has its characteristics and scope of application. It needs to be selected according to the needs of the actual problem. Sometimes, a mixture of several algorithms may be needed.

After selecting the algorithm, choosing a suitable programming language for specific programming work is necessary. There are no unified requirements for which language to choose, which mainly depends on the programmer's preference. C++, Java, Python, and MATLAB are all possible options. Beginners are advised to start with simple languages like Python or MATLAB.

3.   **Letting machines learn from data**

The machine learning process trains the sample data in the selected algorithm, looks for the hidden regularity from the sample data, and finally gets a suitable model. If the output does not match the known results, then the machine should adjust the model parameters and recalculate (i.e., repeat the learning process) to make the match happen.

4.   **Testing and application of the model**

After learning, the model needs to be tested (like an exam after learning). This is the way to evaluate the quality of learning. The test was performed with another part of the sample data besides the training data. If the test result is not ideal, it indicates some problems in the model. There are two possible problems: (1) the sample data might not be representative, and (2) the algorithm might not be appropriate. The possible solutions are: (1) recollect and train the model with new sample data; and (2) choose another algorithm.

After the model tests are passed, the model is ready to work. When you have new input data, the model will automatically make the correct judgment or prediction based on the new data. This ability of a machine to process new data is called generalization ability.

Arthur Samuel developed a checkers program in 1952. Some well-known chess players have lost to this program. Samuel debunked the idea that machines could not surpass humans and created "machine learning through this program." Machine learning does not require deterministic programming to equip machines with skills and learn by themselves like humans.

Machine learning seems simple, but it is very complicated to operate. It requires a large amount of knowledge reserve. The learning results may differ from methods and textbooks like human learning.

(3)   Types of machine learning

There are currently three types of learning: supervised, unsupervised, and reinforcement learning. Different learning methods also represent different machine learning techniques, and their algorithms are also very different.

1.   Supervised learning

Supervised learning, the most common type of learning, is a "teacher (teacher is the label)" type. "Teacher" plays a supervisory role in giving standard answers and correcting mistakes. It can also be understood as learning by using textbooks with answers.

As shown in Fig. 3.7, if a blue ball is an input to the machine, the correct output should also be a blue ball (the answer), then the model obtained is correct. The

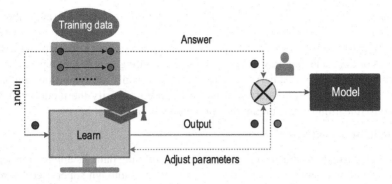

Fig. 3.7 The process of supervised learning

machine returns the error message if the output is a red ball. The program automatically modifies the parameters and continues the calculation until it outputs the correct answer (the blue ball).

Supervised learning is similar to the parameter estimation method in system identification (mathematical model). In this case, "modifying parameters" is "learning." How to modify the parameters will be described in the next section.

Let us observe how children learn about cats and dogs. Without a teacher (adult) telling them which are cats and dogs, it is difficult for a child to tell the difference first. Supervised learning is when a child mistakes a dog for a cat, and a teacher points out mistakes and corrects them. Thinking deeply about the development process of children will be beneficial to understanding and developing machine learning.

2. **Unsupervised learning**

In unsupervised learning, there is no known training data. Instead, the machine directly uses the data for modeling, and it needs to learn by itself to explore the inherent regularities in the data. Unsupervised learning is a self-learning style, which is more complex than supervised learning. Figure 3.8 shows the difference between these two learning styles, using cat and dog recognition as an example.

In recognizing cats and dogs, if there is supervised learning, many images of cats and dogs need to be artificially tagged and turned into training samples. Then, the

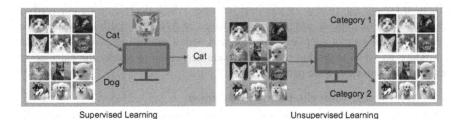

Fig. 3.8 Comparison of supervised and unsupervised learning

machine learns to recognize the cats and dogs; in unsupervised learning, there is no need to mark the pictures of cats and dogs in advance. Instead, the machine can directly distinguish the pictures of cats and dogs, but it does not know which pictures are cats or dogs and can only distinguish them according to different types.

There is also a semi-supervised learning method. Only a tiny portion of a batch of data is used as training data, and the rest is used to verify the model, which is supervised learning. The amount of training data is small.

3. Reinforcement learning

Although unsupervised learning does not require training data, its application scope is limited and is mainly used for data classification. Although supervised learning applies to a wide range, it needs a lot of training data to learn, which is time-consuming and not efficient. Reinforcement learning differs from the above two types of machine learning, which doesn't require training data and learns by doing. Reinforcement learning targets problems related to action and time, such as chess, self-driving cars, process control, market decision-making, etc., which is promising.

Take learning snow skiing as an example. If we fall, the brain will give us bad feedback that the action is wrong. If the action is done correctly, the brain will give us good feedback (reward) that it is the correct skiing move. The skiing learning process is a reinforcement learning process.

Reinforcement learning is a trial and error process in which the machine does something, judges whether the results are good or not, and then adjusts action based on the feedback (i.e., judgments). Through continuous tweaking and trial and error, the machine can choose the actions to achieve the best results.

The core of reinforcement learning is to let machines learn and make decisions by themselves, which is a way of learning in action. This type of learning was developed in the 1950s, but it just became famous in 2016 when AlphaGo beat a top Go player.

Machine learning is a general term for several algorithms widely used in many areas. Our joint efforts on machine learning will result in a bright future in engineering construction.

3.3 Dissect an Intelligent Algorithm

There are many algorithms for machine learning, which deep learning emerged in 2006 and is more popular. Deep learning is based on ANN. Currently, there are more than 30 algorithms for deep learning. We start with a simple regression problem to introduce an ANN algorithm through formula derivation to master the calculation method and understand the nature of AI.

(1) General algorithm of a regression problem

Let us start with a simple linear regression problem with only one variable and then compare it with an intelligent algorithm to see the difference. Only 5 points are selected to simplify the analysis, as shown in Fig. 3.9.

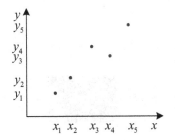 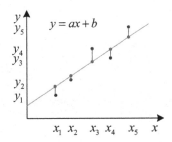

Fig. 3.9 Scatter diagram and fitting diagram

In Fig. 3.9, five groups of data (x_i, y_i) are known, the figure on the left is a scatter graph (a data visualization method), and the figure on the right is a fitting graph using a straight-line fitting as $y = ax + b$. Since coefficients a and b (also known as weights) are unknown, the two coefficients (weights) need to be calculated according to the known data (i.e., the training samples of machine learning), which is generally carried out using the least square method.

The above regression problem has a precise mathematical model $(y = ax + b)$. As long as the weights a and b are determined, y can be predicted according to x. The program for this problem is simple to write. You have these five data sets if you don't have a mathematical model. How do you make predictions based on new data? Machine learning should solve this kind of problem using intelligent algorithms.

(2) Intelligent algorithm of a regression problem

If the intelligent algorithm is used for the same five data sets in Fig. 3.9, there is no need to pre-determine the relationship between data points (straight line or curve). The relationship can be determined during the machine learning process, which is the advantage of intelligent algorithms.

First, select an intelligent algorithm (for example, ANN), and then use these five data sets for training, let the machine learn to find the regularity in the data. When x_1 is input, y_1 should be obtained. If the result is not y_1, the machine will automatically adjust the parameters until the result becomes y_1 (this process is supervised learning). After all data training (i.e., machine learning) is completed, the machine will automatically establish a predictive model (unknown to the users). When x_k is input into this model, the machine will automatically output the correct y_k, as shown in Fig. 3.10.

The conventional algorithm needs to determine the mathematical model in advance for the same regression problem and then solve for the model's parameters. An intelligent algorithm requires only enough data, leaving the task of finding appropriate functions for the machine. So how does the machine work? The following analysis of the ANN algorithm explains the mechanism.

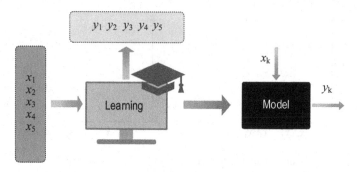

Fig. 3.10 A predictive model of intelligent algorithm

(3) **How neurons work—biological implications**

Artificial Neural Network (ANN) was proposed in 1943 and has experienced a tortuous course of development. To understand how ANN works, it is necessary to know something about neural networks in biology.

The actual neural network structure seems complicated. It is simplified into a schematic diagram to facilitate understanding, as shown in Fig. 3.11. A neuron consists of a cell body, multiple dendrites, an axon, and multiple synapses at the end of the axon. The dendrites, which receive signals (also known as information) from synapses in multiple other neurons, are the input; The axon sends signals to the dendrites of several other neurons through the synapse as the output end. One neuron can connect to thousands of others, and the human brain has approximately 100 billion neurons.

A neuron has two states, as shown in Fig. 3.12. When a neuron receives a signal, it adds up the input and compares it to the neuron's intrinsic "threshold." If the sum of the input signals is equal to or greater than the threshold value, the neuron is activated (in an "active" state) to transmit signals to other neurons. Otherwise, when the sum of the input signals is smaller than the threshold value, the neuron is in an "inactive" state and unable to send signals to other neurons.

Above is a brief introduction to biological neural networks. The following text involving neurons borrows the terms (even if you do not understand them, it does not affect understanding the context). The human brain has no direct connection with

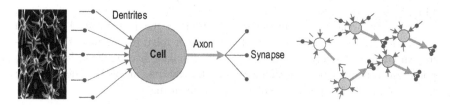

Fig. 3.11 Schematic diagram of the neuron

Fig. 3.12 Two states of a neuron

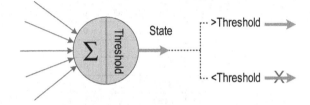

Fig. 3.13 Neuron model

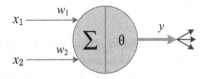

intelligent algorithms, just mathematical models. However, the ANN was inspired by biology. Such inspiration among disciplines is the source of innovation.

According to the neuron structure, a mathematical model can express this idea: the artificial neuron, or neuron or element for short. The neuron model parameters include inputs x_1, x_2, the output y, and the threshold θ. The connection strengths between the neuron and the inputs are w_1 *and* w_2, the weights or connection coefficients (i.e., correction coefficient), as shown in Fig. 3.13.

The neuronal model shown in Fig. 3.13, also known as a Perceptron, is not hardware but just a mathematical model. The model works as follows:

$$\begin{cases} z = [(x_1 w_1 + x_2 w_2) - \theta] \geq 0, & y = 1 \\ z = [(x_1 w_1 + x_2 w_2) - \theta] < 0, & y = 0 \end{cases} \tag{3.1}$$

where z is called the weighted input, $y = 1$ means active, $y = 0$ represents inactive. This model can be used for classification problems.

Figure 3.14 shows an example, where $w_1 = 5$, $w_2 = 2$, $\theta = 20$. According to the image, when $z \geq 0$, $y = 1$, there is signal output, and neurons are activated. When $z < 0$, $y = 0$, there is no signal output, and the neuron is not activated. We call the function represented by the graph in Fig. 3.14 the element step function. Therefore, the input–output relationship of the neuron can be expressed by the element step function.

If $z = [(x_1 w_1 + x_2 w_2) - \theta]$ is the neuron's weighted input, then the input–output relationship can be expressed as follows.

$$y = u(z) = \begin{cases} 1(z \geq 0) \\ 0(z < 0) \end{cases} \tag{3.2}$$

Equation (3.2) is called the activation function of the neuron, also known as the transfer function or the excitation function. It will be generally expressed as $h(z)$,

x_1	x_2	z	Act	y
2	3	$-4<0$	×	0
3	5	$5>0$	√	1
0	1	$-18<0$	×	0
3	4	$3>0$	√	1
6	0	$10>0$	√	1

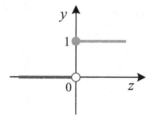

Fig. 3.14 An example neuron

which is used to activate the weighted input signal of the neuron. The activation function is an essential concept.

The above "0–1" function has only two values suitable for the classification problem.$(x_1 w_1 + x_2 w_2) = \theta$ is called the dividing line. To be convenient, we can replace the threshold value θ in the z expression with another symbol b. Let $b = -\theta$, then we can get:

$$z = x_1 w_1 + x_2 w_2 + b \tag{3.3}$$

where b is called bias which controls how easily a neuron can be activated, which can also be treated as a weight of an input with a value of 1 (b is usually a negative number). The advantage of this treatment is that the expression in Eq. (3.3) is unified and can be expressed as a matrix with multiple inputs, which is convenient for programming.

A continuous activation function is often needed to representing the change of "active degree." The typical one is Sigmoid, called logistic regression function, which can be used for classification and regression problems, and its expression is as follows.

$$S(z) = \frac{1}{1 + e^{-z}} = \frac{1}{1 + \exp(-z)} \tag{3.4}$$

The graph of the Sigmoid function is shown in Fig. 3.15. This is the most commonly used activation function in ANN. The purpose of introducing activation functions is to increase the nonlinearity of ANN.

The Sigmoid function is characteristically continuous differentiable and nonlinear. When the output value is close to 1, it indicates that the neuron activity is high and vice versa, and thus some parameters with continuous values can be simulated. There are also many other forms of activation functions.

The working process of each neuron can be divided into two operational steps: summation and activation. The following is a summary of the working process of the neuron model, which will be the basis of ANN and deep learning, and needs to be well understood.

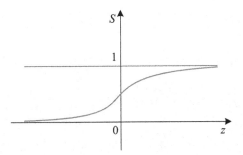

Fig. 3.15 Sigmoid activation function

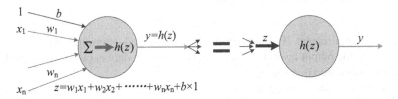

Fig. 3.16 The working process of a neuron

Figure 3.16 shows how the neuron works. A neuron has n inputs $(x_1, x_2, \ldots x_n)$. First, multiply these inputs with the corresponding weights, sum them up (Σ) to get the linear expression z of the weighted inputs, and then put z into the activation function to get the output $y = h(z)$. In fact, "summation and activation" represent "linear and nonlinear" operations, respectively.

Note that there is only one output from each neuron. Still, it can be transmitted simultaneously to different neurons as input, and different neurons could modify this input differently.

In addition, a neuron can be treated as a system with an input of z, a system function of $h(z)$, and an output of y. Since $h(z)$ is a nonlinear function, the system is also nonlinear. If you have many neurons connected, the nonlinear interactions between the neurons will be robust, and you will have a complex network system.

Above in this section is how a neuron works. Although the calculation of a single neuron is not complicated, the connection between neurons will become very close and complex after many neurons form a network system. The calculation results will affect each other and reflect the output results.

(4) **ANN's working process (positive question)**

An artificial neural network (ANN) is a network that consists of multiple connected neurons (elements for short). The structure of ANN has been preliminarily introduced in Sect. 3.1 (see Figs. 3.2 and 3.3).

Typical ANN structure consists of three layers: input, hidden, and output. The number of input data determines the number of elements in the input layer. The number of elements in the hidden layer depends on the body condition, and there is

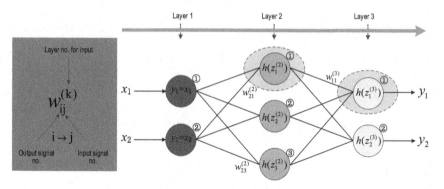

Fig. 3.17 Composition and numbering of an ANN

no unified requirement. The number of answers needed determines the number of elements in the output layer. Except for the input layer, data in other layers needs to be transformed into non-linear.

Now let's construct a neural network (also known as a fully connected network), as shown in Fig. 3.17, and analyze its working process by taking element ① in the hidden layer (layer 2) and output layer (layer 3) as an example. As you can see, ANN also follows the two steps of "summation and activation."

Figure 3.17 shows that ANN consists of 2 input elements, three hidden elements, and two output elements. The activation function adopts the Sigmoid function, and the data is transferred from the input end to the output end. In addition, it is generally assumed that the activation function is the same for all elements (the function could be different, but the calculation would become very complicated).

Because there are many elements, the sequence of various symbols needs to be specified. The layer number of receiving signal is placed in the upper corner, and the element sequence number is placed in the lower corner. The first lower subscript is the element number connected to the previous layer (the element that outputs the signal). The second lower subscript is the element number of the current layer (the element that receives the signal), as shown in Fig. 3.17. Some books may place the numbers in a reserve order (i.e., the first lower subscript is the element number for the current layer while the second subscript is the element number of the previous layer), and you may need to pay attention to avoid misunderstanding.

The symbolic specification is essential. If there are too many elements, unnumbered elements will confuse. Readers may not understand some of ANN's books because they could not understand the symbols. Figure 3.18 shows the representation of each symbol of element i in layer k (except the input layer), which is an alternative representation of Fig. 3.17. It is necessary to master this representation method to build a firm foundation for subsequent derivation and programming.

After the construction of ANN, we will start to analyze the whole process of a signal (data) transferring from the input end to the output end of ANN.

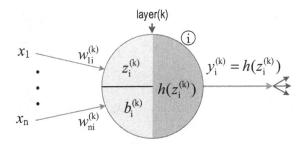

Fig. 3.18 General representations of neural network symbols

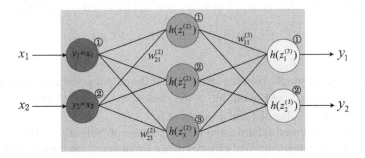

Fig. 3.19 Neural network represented by the system approach

Suppose you are still not used to the above representation in biological terms. You can discard the above terminology and use the system approach described in Chap. 2, as shown in Fig. 3.19. Each neuron is an element of the system, and the connections form the system structure. The system model comprises a set of complex and nonlinear mathematical expressions. The parameters to be determined are multiple weights (including bias).

Looking at ANN from the system's point of view will help us treat the machine learning algorithm correctly and break down the mystery. However, the original physiology description of the brain's physiological structure did inspire researchers' thinking and played a decisive role in creating this algorithm. This is innovation, which is very important!

1. **Input Layer (Layer 1)**

The input layer is responsible for reading the data and output as-is. Figure 3.20 shows 2 neural elements corresponding to 2 inputs (x_1, x_2). The task of the element is to pass the input data to the elements of the hidden layer. For instance, element ① passes the x_1 to elements ①, ②, and ③ in the hidden layer.

2. **Hidden Layer (Layer 2)**

The hidden layer is the core of ANN, which is responsible for linear and nonlinear data processing. First, the element performs a linear summation. The three elements of the hidden layer (Layer 2) receive signals from each element of the input layer,

Fig. 3.20 Input and output
relationship of the input
layer (layer 1)

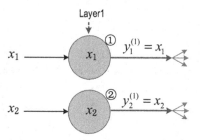

respectively, and each element has a different weight correction for different inputs.
Taking element ① as an example, there are two inputs, then the weighted input is:

$$z_1^{(2)} = w_{11}^{(2)} y_1^{(1)} + w_{21}^{(2)} y_2^{(1)} + b_1^{(2)} \times 1 \qquad (3.5a)$$

where, $w_{21}^{(2)}$ is the weight between element ② of the output signal and the receiving
element ① of layer 2, and so on; $b_1^{(2)}$ is the element ① of layer 2's bias and can be
treated as an input value of 1 and a weight of $b_1^{(2)}$.

Second, the neurons perform a non-linear operation of "activating." The weighted
summation of the input signal of element ① is a linear operation, which needs to be
put into the activation function h(z) for nonlinear conversion. The output of element
① is as follows:

$$y_1^{(2)} = h(z_1^{(2)}) = h(w_{11}^{(2)} y_1^{(1)} + w_{21}^{(2)} y_2^{(1)} + b_1^{(2)} \times 1) \qquad (3.5b)$$

Figure 3.21 graphically shows the relationship between the input and output of
element ①, and the results of summation and activation. Similarly, one can obtain the
output of several other elements. All these outputs are passed to layer 3 (the output
layer).

Note that, due to the different weights of the connections, the output signal to the
elements in the next layer will still be corrected. For example, element ① transmits
the output $y_1^{(2)}$ to element ① and element ② of layer 3, respectively, but the input

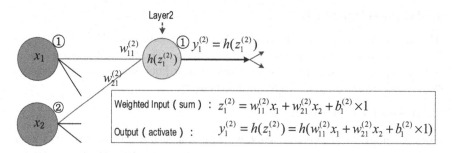

Fig. 3.21 Relationship between input and output of the hidden layer (Layer 2)

Fig. 3.22 Correction of the input signal by receiving element

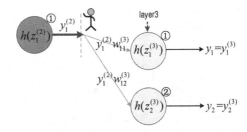

of the receiving element ① of layer 3 is $w_{11}^{(3)} y_1^{(2)}$ not $y_1^{(2)}$, and element ② of layer 3 receives $w_{12}^{(3)} y_1^{(2)}$ not $y_1^{(2)}$. This is the correction effect of weights, as shown in Fig. 3.22.

3. Output Layer (Layer 3)

The output layer is responsible for the output of the entire network. The operation of each element in data processing is similar to that of the hidden layer. It is called "summation and activation." Take element ① as an example to illustrate the whole process. Element ① receives signals from 3 elements of the second layer (the hidden layer), so there are three inputs, and its weighted inputs are

$$z_1^{(3)} = (w_{11}^{(3)} y_1^{(2)} + w_{21}^{(3)} y_2^{(2)} + w_{31}^{(3)} y_3^{(2)} + b_1^{(3)} \times 1) \tag{3.6a}$$

Input the above calculation result (weighted input) into the activation function $h(z)$ for activation operation, and the output of element ① of layer 3 can be obtained as

$$y_1 = y_1^{(3)} = h(z_1^{(3)}) = h(w_{11}^{(3)} y_1^{(2)} + w_{21}^{(3)} y_2^{(2)} + w_{31}^{(3)} y_3^{(2)} + b_1^{(3)} \times 1) \tag{3.6b}$$

Figure 3.23 shows the relationship between the inputs and outputs of element ① of layer 3 and the results of summation and activation. In the same way, we can get the output of the second element, which completes the whole calculation process.

It can be seen that the output of each element in layer 3 contains the comprehensive effects of all relevant elements in the previous two layers (the hidden layer and input layer), including linear operation (weighted summation) and nonlinear operation (activation). ANN is such a nonlinear coupled structure, affecting the whole body.

The signal transmission problem of ANN is described above. Except for the input layer, the signal transmission process of each element in other layers is the same, which is carried out following the steps of "summation and activation." The calculation result of the activation function is the element's output, which is then passed to the next layer's relevant elements, as shown in Fig. 3.24. $k = 2, 3$ represents the hidden layer (layer 2) and the output layer (layer 3), respectively.

From the system's perspective, ANN is a system, and neurons are system elements. The relationship between input and output determines the correlation between

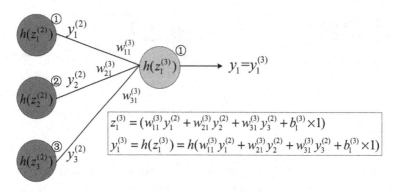

Fig. 3.23 Relationship between input and output of output layer (Layer 3)

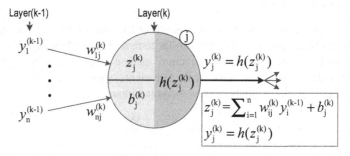

Fig. 3.24 Two operations for each element

elements, and the weight represents the strength of the correlation. ANN is a complex nonlinear dynamic system requiring more learning to understand fully.

Matrix representation of neural network signal transmission (readers who are unfamiliar with matrices may skip this section).It can be found from the above derivation that if the neural network has many elements, the above operation would be very complicated. To simplify computer programming, matrix expression is a better option. Matrix or linear algebra is the primary language of modern mathematics, and its essence is not complicated, but many people get stuck with the terminologies and symbols. The matrix form of the above operations is given below.

According to Fig. 3.18, for the hidden layer (layer 2), the input matrix is the output matrix Y of the input layer (layer 1)$^{(1)}$, the weighted input (summation) matrix is $Z^{(2)}$, the output matrix is $Y^{(2)}$, the weight matrix is $W^{(2)}$, the bias matrix is $B^{(2)}$, and the expressions are as follows:

$$Y^{(1)} = \begin{bmatrix} x_1 \\ x_2 \\ x_3 \end{bmatrix} \quad Z^{(2)} = \begin{bmatrix} z_1^{(2)} \\ z_2^{(2)} \\ z_3^{(3)} \end{bmatrix} \quad Y^{(2)} = \begin{bmatrix} y_1^{(2)} \\ y_2^{(2)} \\ y_3^{(2)} \end{bmatrix}$$

$$W^{(2)} = \begin{bmatrix} w_{11}\,w_{12}\,w_{13} \\ w_{21}\,w_{22}\,w_{23} \\ w_{31}\,w_{32}\,w_{33} \end{bmatrix}^T \quad B^{(2)} = \begin{bmatrix} b_1^{(2)} \\ b_2^{(2)} \\ b_3^{(2)} \end{bmatrix}$$

According to the operation rules of matrix multiplication, the weighted summation matrix of the hidden layer (layer 2) is:

$$Z^{(2)} = W^{(2)}Y^{(1)} + B^{(2)} \tag{3.7}$$

The output of the hidden layer is:

$$Y^{(2)} = h(Z^{(2)}) \tag{3.8}$$

A matrix can also represent the output layer (layer 3) in the same way as for layer 2. Referring to the representation of layer 2, the matrix of layer three can be written as:

$$Z^{(3)} = W^{(3)}Y^{(2)} + B^{(3)}$$
$$Y = Y^{(3)} = h(Z^{(3)}) \tag{3.9}$$

The matrices in the formula are as follows:

$$Z^{(3)} = \begin{bmatrix} z_1^{(3)} \\ z_2^{(3)} \end{bmatrix} \quad Y^{(3)} = \begin{bmatrix} y_1^{(3)} \\ y_2^{(3)} \end{bmatrix} \quad W^{(3)} = \begin{bmatrix} w_{11}\,w_{12}\,w_{13} \\ w_{21}\,w_{22}\,w_{23} \end{bmatrix}^T \quad B^{(3)} = \begin{bmatrix} b_1^{(3)} \\ b_2^{(3)} \end{bmatrix}$$

It can be seen that the working process of neural networks expressed in the form of a matrix is more straightforward and clear, which is incredibly convenient for programming.

From the above derivation, we can summarize the general formula. For layer k, according to the steps of "summation and activation," the weighted input matrix and output matrix can be written in the following form.

$$Z^{(k)} = W^{(k)}Y^{(k-1)} + B^{(k)}$$
$$Y^{(k)} = h(Z^{(k)}) \tag{3.10}$$

where $k = 2, 3$ represents the hidden layer (layer 2) and the output layer (layer 3). The expression content of each matrix in the formula needs to be determined according to the specific situation. The specific expression can be easily written as long as the symbol rules are remembered. For example, for the weight matrix $W^{(k)}$, each element in line i is the corresponding weight of element i in layer k.

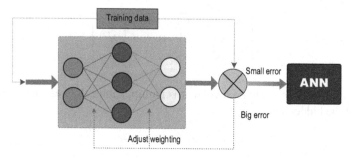

Fig. 3.25 The training (learning) process of ANN

Equation (3.10) shows the relationship between the previous and the current layers. After the weighted correction and summation operation, the output of the previous layer becomes the weighted input of the current layer; through the action of the activation function again, it becomes the output of the current layer (i.e., the input of the next layer). This form of signal propagation is also applicable to multi-layer neural networks.

(5) ANN's learning process (inverse problem)

Once the ANN system is built, the next task is to train the network to uncover hidden regularities in the data that cannot be expressed in explicit mathematical formulas and develop a model to apply to other data. Figure 3.25 shows the training process of ANN, that is, the learning process.

The training data is first input into the ANN model, then the calculation is carried out according to the abovementioned steps. Finally, the output of ANN is obtained. Suppose the output (calculated value) is inconsistent (i.e., the error is not acceptable) with the known values (actual values). In that case, ANN will modify weights automatically and repeat the calculation according to the above steps (also called iterative operation) until the output is consistent with the known result (i.e., the error is acceptable). After this training process is done, a reasonable ANN model is obtained. The error acceptance judgment and weight adjustment will be described in detail below.

In machine learning, the process of using sample data to determine model parameters (such as weights) is known as learning, while it is called parameter estimation (or identification) in mathematics. It is a kind of "inverse problem."

The ANN training process looks very similar to the parameter estimation process in the system identification. They both minimize the error between the calculated and the actual values by adjusting the model parameters. However, system identification has a transparent, solvable mathematical model, while ANN (machine learning) only has data.

The learning process of ANN mentioned above is typical supervised learning, which needs robust sample data as support. Of course, the trained ANN must have the generalization ability to adapt to new data, which is the ultimate goal of machine learning.

(6) Method of adjusting the weight

The key of ANN is to determine network structure parameters, namely weight, which is a multi-iteration calculation process. Its updating principle is "new weight = old weight + change." The critical problem is how to represent the change. This aspect involves mathematical optimization problems and requires familiarity with concepts such as derivatives, especially derivatives of complex functions (chain rule), slope, and gradient. The backpropagation (BP) method is briefly introduced here, typically used to represent the change. The network using this method is also called BP neural network.

1. Error expression—error function

The error is the difference between the output value of ANN (the calculated value) and the actual value (the expected value). In ANN, the "error function" represents the error, also called the loss function. Generally, the sum of squares of the output errors of each element in the output layer is used.

$$E = \frac{1}{2} \sum_{i=1}^{m} e_i^2 = \frac{1}{2} \sum_{i=1}^{m} (y_i - t_i)^2 \tag{3.11}$$

where m represents the number of elements in the output layer; y_i is the output value of element i, t_i is the corresponding actual value; e_i is the output error of element i.

We define the error as $e_i = (y_i - t_i)$ or $e_i = (t_i - y_i)$, which does not affect the following calculation results. The 1/2 appears in Eq. (3.11) to eliminate the coefficients in the derivation of the formula.

In Eq. (3.11), t_i is the given number, $y_{i,}$ involves the weight of the related elements of each layer. The weight w can be regarded as a variable of E's error function and expressed as $E(w)$. This is a function with many variables (weights). One of the significant tasks for ANN is to adjust the weight according to $E(w)$, which is also a learning task.

2. **Gradient descent**

Theoretically, when the output value of ANN is equal to the actual value, the corresponding weight is the best value we are looking for. But the error is always there, and it's not realistic to get $E(w) = 0$. To solve this problem, $E(w)$ can be required to reach a specified small value, where the weight is the optimal estimate. This is a parametric optimization problem in math.

According to mathematical principle, for $E(w)$, as long as the derivative is 0, $E(w)$ has the extreme value (minimum value), and the corresponding weight is the optimal value. Since there are multiple weights, we need to use partial derivatives.

$$\frac{\partial E(w)}{\partial w_{ij}} = 0 \rightarrow w_{ij} \tag{3.12a}$$

The above expression is a general expression representing the derivative of $E(w)$ for all weights. Equation (3.12a) can be generalized by combining each derivative as its component into a vector form, as shown in Eq. (3.12b). This vector is called the $E(w)$ gradient at points $w_{11}, w_{21}, \ldots, b_1, \ldots$

$$\left(\frac{\partial E}{\partial w_{11}}, \frac{\partial E}{\partial w_{21}}, \ldots, \frac{\partial E}{\partial b_1}, \ldots \right) \qquad (3.12b)$$

These components are the familiar concept of slope (gradient is a vector, the slope is a scalar). For a variable, gradient and slope are not strictly distinguished. We can use K_{ij} substituting partial derivatives for readers who are not used to the partial derivative notation. For example, K_{12} is the slope of weight w_{12}.

If we can solve the derivative of $E(w)$ analytically, we can solve the weights simultaneously. Unfortunately, with the increase in the number of ANN elements, $E(w)$ expression will become more and more complex, and it is challenging to find the derivative by analytical method. In this case, only numerical methods can be used to gradually approximate the minimum value of $E(w)$, which is the gradient descent method in optimization theory. The optimal value is the weight corresponding to the minimum $E(w)$.

Gradient descent is like a person walking from the top to the foot of a mountain with a flashlight at night (Fig. 3.26). If one doesn't know the route down the hill, one has to find his way roughly in the downhill (gradient) direction, deciding which direction is the steepest at each step. By constantly adjusting the direction and step length and correcting the descent route, we will eventually reach the foot of the mountain (the lowest point). The gradient here is the slope of the hill, and the step length is η. The best way is to go downhill, always along the steepest direction.

Gradient descent is a numerical method. The same way a person would descend a mountain, descending in the direction of the maximum gradient, constantly adjusts the step size a little bit closer to the lowest position. Since the composition of $E(w)$ is very complex, this approximation method should be more practical if an analytical solution cannot be carried out. It does not require derivatives, and a simplified iterative solution can be used to find the optimal weight.

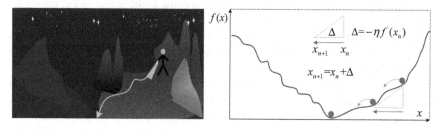

Fig. 3.26 Gradient descent corresponds to the exploration process of descending the mountain

Fig. 3.27 Relationship
between slope and increment

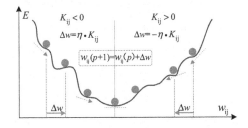

3. Weight update formula

The so-called update weight is the change made based on the last (pth step) weight value so that $E(w)$ is closer to the minimum value. According to the updating formula created by the gradient descent method, the updating amount of weight can be obtained as follows:

$$w_{ij}(p + 1) = w_{ij}(p) - \eta \cdot \frac{\partial E(w)}{\partial w_{ij}} = w_{ij}(p) - \eta \cdot K_{ij} \qquad (3.13)$$

where η is the learning rate, which plays a role in regulating the weight size, like the "step length" of a person walking down from the hilltop, which is generally $0 < \eta < 1$. The weight change Δw is directly related to the weight gradient (slope K_{ij}). After calculating the slope of the weight, a gradient in the form of Eq. (3.12b) can be obtained. The gradient descent method searches for the optimal weight along the direction of the maximum change of the gradient.

Regarding Eq. (3.13), if $K_{ij} > 0$, you have to decrease w_{ij}, w_{ij} will move to the left, use minus sign; If $K_{ij} < 0$, we need to increase w_{ij}, w_{ij} will move to the right, use plus sign, as shown in Fig. 3.27. The magnitude of movement of w_{ij} is controlled and regulated by η.

4. Calculation of the slope

(a) Slopes of weights in the output layer (layer 3)

According to the weight update Eq. (3.13), the critical problem is determining the slope. Based on the principle from particular to general, the element ① of the output layer (layer 3) in Fig. 3.28 is taken as an example for derivation. Then, the general expression is given.

As shown in Fig. 3.28, there are three signals transmitted from the hidden layer (layer 2). With input in the form of bias, it can be considered that there are four inputs in total, corresponding to four weights. For the output layer element ①, $E(w)$ can be treated as a composite function of $z_1^{(3)}$ ($E \rightarrow z \rightarrow w$) whose intermediate variable is the weighted input. Derivative of $E(w)$ concerning each weight can be written according to the chain rule of the composite function. Taking $w_{11}^{(3)}$ as an example, the expression of its derivative is as follows.

Fig. 3.28 Calculation of
element ① in the output layer
(layer 3)

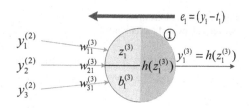

$$\frac{\partial E(w)}{\partial w_{11}^{(3)}} = \frac{\partial E(w)}{\partial z_1^{(3)}} \cdot \frac{\partial z_1^{(3)}}{\partial w_{11}^{(3)}} \tag{3.14a}$$

The above formula can be divided into two parts. The first part is the derivative of $E(w)$ concerning the weighted input, which is defined as the weighted input error of the neuron and is denoted as $\delta_1^{(3)}$. The second part is the derivative of weighted input concerning weight, which is derived as follows.

$$\frac{\partial z_1^{(3)}}{\partial w_{11}^{(3)}} = \frac{\partial}{\partial w_{11}^{(3)}} (w_{11}^{(3)} y_1^{(2)} + w_{21}^{(3)} y_2^{(2)} + w_{31}^{(3)} y_3^{(2)} + b_1^{(3)} \times 1) = y_1^{(2)}$$

Thus,

$$K_{11}^{(3)} = \frac{\partial E(w)}{\partial w_{11}^{(3)}} = \delta_1^{(3)} \cdot y_1^{(2)} \tag{3.14b}$$

And we can do the same thing for a couple of other slope expressions.

$$K_{21}^{(3)} = \delta_1^{(3)} \cdot y_2^{(2)}, \ K_{31}^{(3)} = \delta_1^{(3)} \cdot y_3^{(2)}, \ \frac{\partial E(w)}{\partial b_1^{(3)}} = \delta_1^{(3)} \cdot 1 \tag{3.14c}$$

It can be seen that the slope of each weight is the product of the weighted input error and the corresponding input. Since the output of the elements in layer 2 is known, the key is to determine $\delta_1^{(3)}$.

According to Fig. 3.28, the output of element ① is $y_1^{(3)} = h[z_1^{(3)}]$, $E(w)$ can be regarded as the compound function ($y_1^{(3)}$) with the intermediate variable as ($E \to y \to z$), and the derivation is still carried out following the chain derivative rule of the compound function.

$$\delta_1^{(3)} = \frac{\partial E(w)}{\partial z_1^{(3)}} = \frac{\partial E(w)}{\partial y_1^{(3)}} \cdot \frac{\partial y_1^{(3)}}{\partial z_1^{(3)}}$$

$$= \frac{1}{2} \frac{\partial [(y_1^{(3)} - t_1)^2 + (y_2^{(3)} - t_2)^2]}{\partial y_1^{(3)}} \cdot \frac{\partial h(z)}{\partial z}\bigg|_{z=z_1^{(3)}} \tag{3.15}$$

It can be simplified as follows:

$$\delta_1^{(3)} = e_1 \cdot h'[z_1^{(3)}] \tag{3.16}$$

Similarly, the weighted input error of element ② can be obtained as $\delta_2^{(3)} = e_2 \cdot h'[z_2^{(3)}]$. The general formula for the weighted input error is as follows.

$$\delta_j^{(3)} = e_j \cdot h'[z_j^{(3)}] \tag{3.17}$$

where j is the sequence number of the element. It can be deduced that the general formula of each weight slope of the output layer (Layer 3) is as follows:

$$K_{ij}^{(3)} = \frac{\partial E(w)}{\partial w_{ij}^{(3)}} = \delta_j^{(3)} \cdot y_i^{(2)} = e_j \cdot h'[z_j^{(3)}] \cdot y_i^{(2)} \tag{3.18}$$

where i represents the element number of input information (layer 2); j represents the element number that receives the information (layer 3).

When the Sigmoid function is selected as the activation function, that is, $h(z) = S(z)$, the result can be obtained as $S'(z) = S(z)[1 - S(z)]$. Substitute this result into Eq. (3.18) to calculate the slopes.

Figure 3.29 shows the error backpropagation process and slope formula of element ①. Through the product of e_1 and $h'[z_1^{(3)}]$, the output error e_1 is transmitted back to the input layer and then multiply it by the corresponding input to get the slope.

Above, the slope expression of the error function on each weight of the output layer (layer 3) is deduced, and the process is straightforward. If written in words, it can be written as follows.

$$\text{Slope} = \text{Weighted input error} \times \text{Error}$$

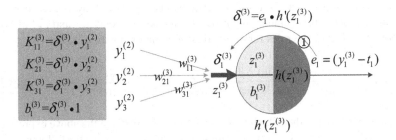

Fig. 3.29 The output error of element ① transmitted back to the input

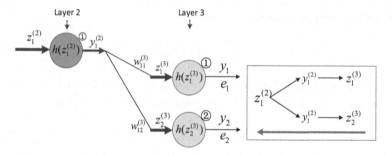

Fig. 3.30 Backpropagation error from the output layer to the hidden layer

After the slope of each weight of the output layer (layer 3) is calculated, the slope of each weight of the hidden layer (layer 2) can be continued to be solved. The idea is to take the results of layer 3 and then deduce the results of layer 2.

(b) Slopes of each weight of the hidden layer (layer 2).

For the hidden layer (layer 2), the slope expression of weight is similar to Eq. (3.18), but the subscript should be replaced with the second layer. The following derivation is made with the example of element ① of layer 2. The derivative of the error function for the $w_{11}^{(2)}$ is expressed as follows.

$$\frac{\partial E(w)}{\partial w_{11}^{(2)}} = \frac{\partial E(w)}{\partial z_1^{(2)}} \cdot \frac{\partial z_1^{(2)}}{\partial w_{11}^{(2)}} = \delta_1^{(2)} \cdot y_1^{(1)} \tag{3.19}$$

The fundamental problem remains to determine the weighted input error of element ① $\delta_1^{(2)}$. For the convenience of derivation, the relationship between the weighted input of element ① $z_1^{(2)}$ and the input of each element in layer three is shown in Fig. 3.30.

According to Fig. 3.30, the weighted input of element ① $z_1^{(2)}$ is transformed by the activation function and transmitted to elements ① and ② of layer 3, respectively, which is a part of the weighted input sum of the two elements $z_1^{(3)}$ and $z_2^{(3)}$ of layer 3. Now let's follow the rules for derivatives of complex functions to derive.

$$\delta_1^{(2)} = \frac{\partial E(w)}{\partial z_1^{(2)}} = \frac{\partial E(w)}{\partial z_1^{(3)}} \cdot \frac{\partial z_1^{(3)}}{\partial y_1^{(2)}} \cdot \frac{\partial y_1^{(2)}}{\partial z_1^{(2)}} + \frac{\partial E(w)}{\partial z_2^{(3)}} \cdot \frac{\partial z_2^{(3)}}{\partial y_1^{(2)}} \cdot \frac{\partial y_1^{(2)}}{\partial z_1^{(2)}} \tag{3.20}$$

Carefully observe the meaning of each part and symbols of the above expression, combined with the previous given various expressions, we can get the following form after derivation.

$$\frac{\partial E(w)}{\partial z_1^{(3)}} = \delta_1^{(3)}, \; \frac{\partial z_1^{(3)}}{\partial y_1^{(2)}} = w_{11}^{(3)}, \; \frac{\partial E(w)}{\partial z_2^{(3)}} = \delta_2^{(3)}, \; \frac{\partial z_2^{(3)}}{\partial y_1^{(2)}} = w_{12}^{(3)}, \; \frac{\partial y_1^{(2)}}{\partial z_1^{(2)}} = h'[z_1^{(2)}]$$

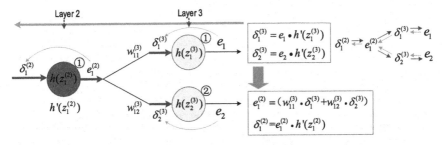

Fig. 3.31 Error backpropagation law from l 3 to layer 2

Therefore,

$$\delta_1^{(2)} = (w_{11}^{(3)} \cdot \delta_1^{(3)} + w_{12}^{(3)} \cdot \delta_2^{(3)}) \cdot h'[z_1^{(2)}] \qquad (3.21)$$

Let $e_1^{(2)} = (w_{11}^{(3)} \cdot \delta_1^{(3)} + w_{12}^{(3)} \cdot \delta_2^{(3)})$, called the output error of element ① in layer 2. One can get:

$$\delta_1^{(2)} = e_1^{(2)} \cdot h'[z_1^{(2)}] \qquad (3.22)$$

It can be seen that, just like the output layer, element ① $\delta_1^{(2)}$ of the hidden layer (layer 2) is the product of output error $e_1^{(2)}$ and derivative of activation function $h'[z_1^{(2)}]$, while $e_1^{(2)}$ is the sum of the product of weighted input error of each element of layer three and its corresponding weight. Figure 3.31 shows the relationship between errors. The error propagates forward from layer 2 to the output of layer 3, that is $e_1^{(2)} \rightarrow e_1^{(3)}, e_2^{(3)}$, the blue arrows in Fig. 3.31 point to the errors.

Substituting Eq. (3.22) into Eq. (3.19), the slope of the weight of element ① $w_{11}^{(2)}$ of layer 2 can be obtained as follows.

$$K_{11}^{(2)} = \frac{\partial E(w)}{\partial w_{11}^{(2)}} = \delta_1^{(2)} \cdot y_1^{(1)} = e_1^{(2)} \cdot h'(z_1^{(2)}) \cdot y_1^{(1)} \qquad (3.23)$$

Compared with Eq. (3.18), they are the same except for the different subscripts. And by the same way, we can derive a couple of other slope expressions.

$$K_{21}^{(2)} = \delta_1^{(2)} \cdot y_2^{(1)} \frac{\partial E(w)}{\partial b_1^{(2)}} = \delta_1^{(2)} \cdot 1 \qquad (3.24)$$

The error backpropagation method avoids the problem of solving the derivative directly, dramatically simplifies the calculation, and plays a vital role in developing ANN.

For the jth element in the k layer, its weighted input error is expressed as follows.

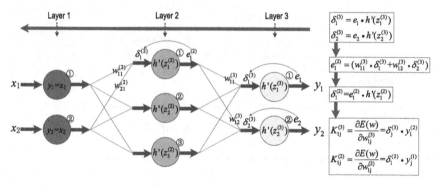

Fig. 3.32 Schematic diagram of error backpropagation law

$$\delta_j^{(k)} = e_j^{(k)} \cdot h'(z_j^{(k)})$$ (3.25a)

The output errors of layer 3 and layer 2 are expressed as follows.

$$e_j^{(3)} = (y_j - t_j)$$
$$e_j^{(2)} = (w_{j1}^{(3)} \cdot \delta_1^{(3)} + w_{j2}^{(3)} \cdot \delta_2^{(3)} + w_{j3}^{(3)} \cdot \delta_3^{(3)} + \cdots)$$ (3.25b)

According to Eq. (3.18), the general expressions of all slope expressions at layers 3 and 2 can be written directly. For element j in the k layer, the slope expression of $E(w)$ for each weight is:

$$K_{ij}^{(k)} = \frac{\partial E(w)}{\partial w_{ij}^{(k)}} = \delta_j^{(k)} \cdot y_i^{(k-1)}$$ (3.26)

Figure 3.32 shows the error e_1 propagating from the output end of layer 3 element ① back to the input end, converted to the output error of layer 2, and then passed to the input end of layer 2.

Substituting Eq. (3.26) into Eq. (3.13), a complete weight update formula (learning rule) can be obtained, which is consistent with Eq. (3.13). Still, the superscript is used to represent each layer, making it clear.

$$w_{ij}^{(k)}(p+1) = w_{ij}^{(k)}(p) + \Delta w_{ij}^{(k)}(p)$$
$$\Delta w_{ij}^{(k)} = -\eta \cdot \delta_j^{(k)} \cdot y_i^{(k-1)}$$ (3.27)

(7) Examples and steps—solving linear regression problem with ANN

For example, let's go back to the linear regression problem mentioned at the beginning of this section, a simple mathematical problem that can be solved easily without ANN. This illustrates the differences between direct solving and ANN.

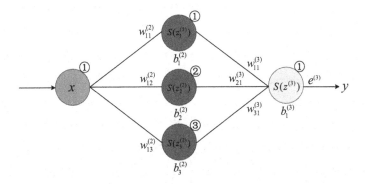

Fig. 3.33 Linear regression problem solved by ANN

As mentioned earlier, five data sets are known as (x_i, y_i), hoping to find a regularity between them (assuming we don't know that the relationship is linear). Now try to set up an ANN structure as shown in Fig. 3.33. The input layer has one element, the hidden layer has three elements, and the output layer has one element. The activation function adopts the Sigmoid function. As to why such a structure should be adopted, there is no standard regulation, and it is generally based on experience.

Following is a brief illustration of the operation process according to the learning (training) steps of ANN. It has to be noted that the real ANN training requires computer programming, and you cannot finish it manually since the amount of calculation is too large.

1. **Network initialization**

Network initialization includes the normalization of training data, the initial weight and bias by random decimal, and the determination of the learning rate.

Firstly, the sample data is normalized. Since the value of the Sigmoid function is between 0 and 1, it is necessary to preprocess the input data and convert it to the range of (0, 1). This processing is called normalization and has been widely used.

Since the Sigmoid function is sensitive to the change of input on the interval [0.1, 0.9] (Fig. 3.15), the training data can be converted to the interval [0.1, 0.9] (mathematically called the mapping). The normalization formula is as follows:

$$s_i = s_{\min} + \frac{(x_i - x_{\min})}{(x_{\max} - x_{\min})}(s_{\max} - s_{\min}) \qquad (3.28a)$$

where x_i is the ith original data; $x_{\max}$, $x_{\min}$ are the maximum and minimum value of the original data, respectively; s_i is the normalized data; $s_{\max}$ and $s_{\min}$ are the corresponding maximum and minimum values, respectively, which can be 0.9 and 0.1, respectively. After the calculation, the data should be converted back to the original to solve for x according to Eq. (3.28a). In addition to Eq. (3.28a), the following equation can also be used for normalization.

Table 3.1 Training data

No	1	2	3	4	5
x	0.11	0.32	0.57	0.76	0.85
y	0.21	0.25	0.67	0.66	0.89

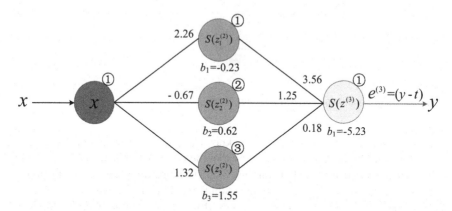

Fig. 3.34 ANN marked with initial parameters

$$s_i = \frac{(x_i - x_{min})}{(x_{max} - x_{min})} \tag{3.28b}$$

Table 3.1 presents the normalized results of the five data groups. It is not enough to rely only on five data sets to train an ANN. A large amount of data is needed to get a better model. Here is only for demonstration purposes.

Secondly, set the learning rate as $\eta = 0.3$. As for setting the initial weight and bias value, it is generally OK to set it randomly, as shown in Fig. 3.34.

According to the characteristics of this example, there are five sets of training data. Still, there is only one element in the input layer, so only one set of data can be calculated at a time, and five times of calculations are needed to complete training. Sum the errors of the five calculations with the slope to get the total error and slope, and then update the weight value.

2. Calculate output and error

Input training data to ANN, such as $x_1 = 0.11$. According to the calculation method of "summation and activation," the output layer's output (calculated value) is finally obtained. The output layer's error e and mean square error E are calculated.

For example, the three hidden layer elements' weighted input, output, and derivative values (layer 2) can be calculated in the following ways.

$$\begin{cases} z_i^{(2)} = w_{1i}^{(2)}x_1 + b_i^{(2)} \times 1 \\ y_i^{(2)} = S(z_i^{(2)}) = \dfrac{1}{1 + \exp[-z_i^{(2)}]} \quad (i = 1, 2, 3) \\ S'[z_i^{(2)}] = S(z_i^{(2)})[1 - S(z_i^{(2)})] \end{cases} \qquad (3.29)$$

3. Calculate the weighted input error

According to the error backpropagation law, starting from the output error of the output layer, the weighted input error δ of each element is calculated layer by layer reversely.

For example, the weighted input error of the input terminals of the three elements in layer 2 is calculated as follows.

$$\delta_i^{(2)} = [w_{1i}^{(3)} \cdot \delta_1^{(3)}] \cdot S'[z_i^{(2)}] \quad (i = 1, 2, 3) \qquad (3.30)$$

4. Calculate the slope

The output layer (layer 3) has one element, three inputs, and one bias, so there are four slopes as calculated by the following formula.

$$K_{1j}^{(3)} = \delta_1^{(3)} \cdot y_j^{(2)} \quad (j = 1, 2, 3) \qquad (3.31)$$

The hidden layer (layer 2) has three elements, each element has one input, and there is one bias, so there are four slopes s, calculated by the following formula.

$$K_{i1}^{(2)} = \delta_i^{(2)} \cdot x_1 \quad (i = 1, 2, 3) \qquad (3.32)$$

At this point, the calculation of the first input data is completed. Then input the second data and repeat Steps 2–4 until the calculation of the fifth input data is completed.

5. Update the weights

The mean square error $E(w)$ was obtained from 5 calculations. Each slope is summed accordingly to obtain the mean square error and each training slope (i.e., $E = \sum E_i, K = \sum K_i$). The weight value is automatically modified according to the weight update formula.

For example, the three hidden layer elements (layer 2), with three weights and three biases, are updated according to the following formula.

$$(w_{i1}^{(2)} - \eta \cdot K_{i1}^{(2)}) \rightarrow w_{i1}^{(2)}, (b_i^{(2)} - \eta \cdot K_{i1}^{(2)}) \rightarrow b_i^{(2)} \qquad (3.33)$$

6. **Iterative operation**

Update the parameters of the entire network based on the updated weights and biases, and then start the calculation again from the first input. Repeat Steps 2–5 until $E(w)$ is sufficiently small to obtain and test the ANN model.

The above is the learning of 5 sets of training data. If there are new training samples, additional data can be added for new learning to achieve better learning results.

In addition, the number of elements in the hidden layer will affect the accuracy and speed of training. If the number of elements is too large, there will be an "overfitting" problem. ANN only remembers training samples but cannot adapt to new data. ANN also has poor generalization ability. On the contrary, "underfit" will not get the correct prediction results.

After the ANN model is established, it has to pass the test in the testing phase before starting to work—output the predicted value using the new input data. Note that the test data must be strictly independent of the training data.

> If one person looks at this data, would they predict it? The answer should be no. Even the most intelligent person can't make predictions based on a data set without the tools, but a machine can quickly predict it. In some areas (such as data prediction), it is hard for humans to do things that are easy for computers to do; in others (such as identifying cats and dogs), it is the opposite. This phenomenon is worth our deep thinking!

The ANN and the statistical methods solve the problem differently for the linear regression problem. The statistical method has a precise mathematical model, linear operation, and two unknown parameters (a and b). The ANN model only deals with data that contains nonlinear operations and has ten unknown parameters (weight and bias) with multiple superposition operations. Therefore, ANN is not the best choice for such a simple problem.

(8) **Solutions to classification problems**

ANN is beneficial not only for regression but also for classification problems. Classification is more common than regression. Recognition of various images, video (images played at 24 frames per second), speech recognition, etc., are classified problems. There are also many classification problems in engineering projects as well. Classification is also a hot topic in AI research at present.

Using ANN to solve the classification problem, the calculation steps are precisely the same as the regression problem, but the input and output are different. The following image recognition is used as an example for a brief explanation.

1. **The form of image storage in the computer**

To better understand the input form of an image in ANN, it is necessary to understand the storage form in a computer.

The primary element is pixels (dots), whose values are numbers (where digital images come from), representing different colors and brightness levels. An image

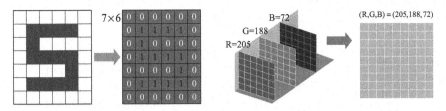

Fig. 3.35 Data form of image in a computer

contains many pixels, usually in the form of resolution. For example, a resolution of 40 × 30 means the image has 1200 pixels (the data dimension is 1200).

Pixel values are stored in a matrix on a computer, as shown in Fig. 3.35. The number in each square is a pixel value, representing a different color.

Figure 3.35 shows a black-and-white image and a color image. For black and white images, the number "0" is used to represent white, and "1" is used to represent black, which is why they are called "binary" images. Image "5" is stored in the computer as a (6 × 7) matrix, which means it has a resolution of 6 × 7 and has 42 pixels.

For color images, a combination of the three primary colors, namely red (R), green (G), and blue (B), is generally used to represent a color. The number represented by each color varies from 0 to 255, indicating the degree of light and shade. Pixel points are represented by (R, G, B). (R, G, B) = (205, 188, 72) in Fig. 3.35, indicating that three primary colors with different degrees of light and shade are combined into orange. Color images in computer storage are composed of a digital three-dimensional array. The length and width of the image resolution and height (thickness) are called the number of channels. The number of channels in the color image is 3, corresponding to three matrices, and each pixel is a combination of the elements of these three matrices. The black and white image has one channel corresponding to one matrix.

2. **Input**

When an image is entered into ANN, the input is the number behind the image. For a binary image (black and white or grayscale), the number of pixels (resolution size) equals the number of input elements. The input elements should be three times the resolution (three channels) for color images. Figure 3.36 is an ANN designed to recognize the numbers "5" and "7"; the image resolution is 7 × 6.

In Fig. 3.36, 42-pixel points in the image are represented as a (7 × 6) matrix. ANN reads pixel values in a specific order, which requires 42 input elements corresponding to pixel values. If it is a color image, such as 1000 × 1000 resolution (non-HD), plus three color channels, then an image will have 3×10^6 data (high-dimensional data), the input element of ANN needs to be set as 3×10^6, the number of weights is much more.

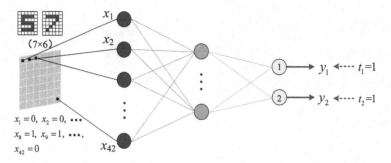

Fig. 3.36 Input and output of image recognition

3. Output

The output values of the regression problem are continuous, while the output values of the classification problem are discrete, belonging to the "either/or." Now take the image of recognizing the numbers "5" and "7" to illustrate. This is printed and easy to recognize but not easy for handwritten forms. Readers can practice by themselves.

Figure 3.36 shows that the image "5" and "7" need to be recognized, respectively, so two output elements should be set. Element 1 corresponds to the number 5, and the correct answer is "1". Element 2 corresponds to the number 7. The correct answer is "1". When using the Sigmoid function as the activation function, the correct output for both element 1 and element 2 should be a number close to 1. If the output is a number close to 0, the recognition result is wrong, and the weight should be adjusted and calculated again.

The computational steps of ANN for image recognition are consistent with the regression problem. According to Eq. (3.11), its error function is carried out. If there are multiple images, the error should be the sum of the errors of all the input images.

Recognizing numbers is a relatively simple classification problem. For more complex images (such as cats and dogs), ANN recognizes them according to some features, but how to extract the features of the image is always tricky. A human can quickly tell a cat from a dog, but whether a machine can do so depends on the suitability of the manually extracted features.

Above, we have analyzed the BP neural network's working process and learning principle in detail. Its essence is a numerical calculation method, and the backpropagation method does not simulate the human brain's learning process.

In principle, three-layer ANN can simulate arbitrarily complex functions. All regression problems can be solved using three-layer ANN, whether linear or nonlinear, unitary or multivariate. Three-layer ANN can also be used if the input is relatively tiny for classification problems. For example, identifying license plate numbers only involves recognizing words, letters, and numbers.

The number of elements and weights will increase dramatically if the input is extensive, making iterative calculation disastrous. Some new features can be added to improve the performance of the ANN, which leads to an increase in the number of hidden layers and the emergence of the mysterious Deep Learning!

3.4 Deep Learning is not a Mystery

Deep Learning (DL) is a general term for a class of multi-layer ANN algorithms, with the number of hidden layers at least more than two and sometimes up to 10 layers, as shown in Fig. 3.37.

Multilayer ANNs have been developed for a long time but have not attracted much attention. Since 2006, when it was given a snazzy new name, "deep learning," the algorithm's popularity has soared. Of course, the name "deep learning" makes the ANN popular since this ANN's hidden layer is "deeper" and closer to the physiological structure of the brain. In mathematics, deep learning is a multi-layer nonlinear network structure that is more complicated than three-layer ANN and is mainly used for classification problems.

Deep learning is one of the hottest topics in AI in recent years, especially after AlphaGo's won the world Go championship in 2016. There is no doubt that deep learning is an excellent machine learning algorithm, which performs better than other algorithms in many fields, but it is not omnipotent. We shouldn't mystify these deep learning algorithms.

In terms of mathematical nature, deep learning is not substantially different from other machine learning methods designed to distinguish (categorize) different objects. But compared to other machine learning, deep learning can capture deeper connections in the data, resulting in more accurate models. At present, a variety of deep learning models have been developed. An example is the convolutional neural network that was applied in image recognition. Another example is recursive neural networks applied in computer vision and language understanding with good results.

Although deep learning is derived from ANN and seems to have only a few more hidden layers, it has gone beyond the ANN framework. The corresponding mathematical processing methods are very different. Figure 3.38 shows the difference

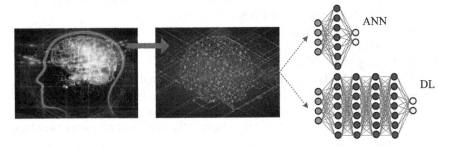

Fig. 3.37 Three and multiple ANNs

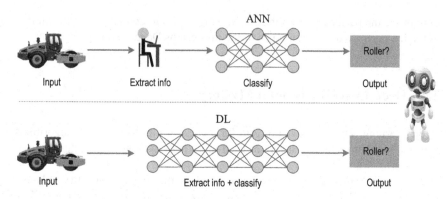

Fig. 3.38 Differences between DL and ANN

between the two types of algorithms in image recognition (roller), which is mainly reflected in feature extraction.

Suppose ANN is used to recognize the roller. In that case, the first step is manually extracting the roller's characteristics. These include the drum and the cabin. These roller's characteristics can be used to train the machine for classification. Humans can tell the difference between a road roller and a car at a glance, but machines cannot necessarily tell the difference with a glance. Therefore, extracting the features that distinguish them from other objects is the key.

ANN requires manual feature extraction, while deep learning is automatic feature extraction by machine. The following is an example of deep learning by using a Convolutional Neural Network for image recognition.

(1) Basic structure of the convolutional neural network

Convolutional Neural Network (CNN) is an algorithm of deep learning. Although CNN was proposed in the 1980s, it emerged with the rise of deep learning. It is also directly related to advances in computer hardware, such as the graphics processor GPU. At present, this algorithm is mainly used for classification problems, especially problems like images with a vast quantity of input and weight. CNN can automatically extract the features of the image and avoid the interference of human factors. Let's first look at the basic structure of CNN.

Due to the massive input in the three-layer ANN image recognition, the first question is whether it is possible to preprocess the image to minimize the calculations in the following stages. That's what CNN is doing.

CNN adds hidden layers with different functions based on three layers of ANN, and two of these hidden layers are essential. The first is a convolutional layer to extract features (such as contours) of images (objects), and the second is a compression layer (it is called the Pooling layer in AI) to compress the images as necessary to reduce the computational load.

Figure 3.39 shows two representations of CNN. Since the number of elements is too large to be drawn directly, multiple planes are generally used to represent each

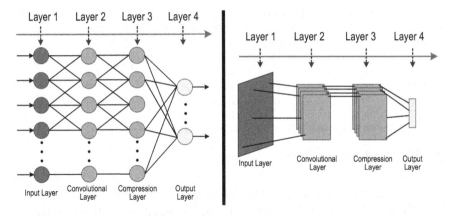

Fig. 3.39 Two representations of CNN

plane, and each plane is occupied with independent elements. The figure on the right is one of the representation methods. Multiple convolutions and compression layers can process the input multiple times.

> In 1958, biologists studying the correspondence between pupil regions and brain neurons in cats discovered that information was processed in stages by different neurons, beginning with the intake of the original signal, such as a balloon. First, some neurons detect objects' edges (contours); Second, the neuron abstracts and determines that the object is round. Finally, after further abstraction, it was determined to be a balloon. CNN was developed with the inspiration of this study.

Initially, CNN was proposed to solve the problem of image recognition. CNN is now widely used in image, video, audio, and text data and has almost become a synonym for deep learning.

(2) **Convolution operation**

The concept of convolution comes from mathematics. It is a mathematical transformation (called operator) that generates the third function y(x), through the operation of two functions, f(x) and g(x), namely:

$$y(x) = f(x) * g(x) = \int_{-\infty}^{\infty} f(t) \cdot g(x - t)\mathrm{d}t \qquad (3.34a)$$

When computing using a computer, the integral becomes the sum, and convolution becomes the sum of two sequences of numbers multiplied by each other over a specific range. Therefore, the above equation can be rewritten as the following equation:

$$y(n) = f(n) * g(n) = \sum_{i=-\infty}^{\infty} f(i) \cdot g(n - i) \qquad (3.34b)$$

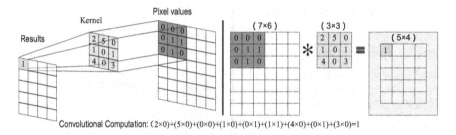

Convolutional Computation: (2×0)+(5×0)+(0×0)+(1×0)+(0×1)+(1×1)+(4×0)+(0×1)+(3×0)=1

Fig. 3.40 Operation of convolution

The above formula still looks a bit complicated (it should be familiar to signal processing engineers), but the actual operation is straightforward. It is an operation between two matrices, defined as the sum of the numerical products of the corresponding positions of the two matrices. Let's see how we can do this in conjunction with image processing.

The image is stored digitally in a computer (Fig. 3.35) and can be treated as a matrix. If you do convolution, you need to create another matrix called the convolution kernel (in signal analysis, it is called a filter). Then, you can use this convolution kernel to convolve with the matrix representing the image. Figure 3.40 shows the specific calculation method.

The convolution operation is a summation operation that uses a small matrix (convolution kernel) to carry out the operation with the original image matrix row by row and column by column to obtain a new matrix representing a new image. The new image contains some main features of the original image, such as shape, contour, etc.

The convolution kernel usually takes a 5×5 or 3×3 matrix, and the numbers in the matrix can be assigned with an initial value (i.e., weight). When doing convolution, the convolution kernel (small matrix) needs to be scanned along with the horizontal and vertical directions of the large matrix, and the corresponding elements are multiplied and then added to get a number, which is used as an element of the new matrix (new image), and so on.

Convolution is a commonly used method in signal analysis, and its essence is a kind of filtering of signals. The image convolution operation filters the image (signal), extracting the image contour features. There is nothing mysterious about it.

When we look at an object, what is the first thing we see (you can close your eyes and then suddenly open them to look at an object, what is the first thing you see?) The answer is the shape of the object! For convolution operation, the purpose is to get the appearance of objects in the image features, namely, edge recognition. Firstly, it finds the "edge" and "outline" of objects. The "edge" and "outline" relate to the adjacent pixels. The image scanning and convolution computation can recognize the edge using the little matrix of the convolution kernels.

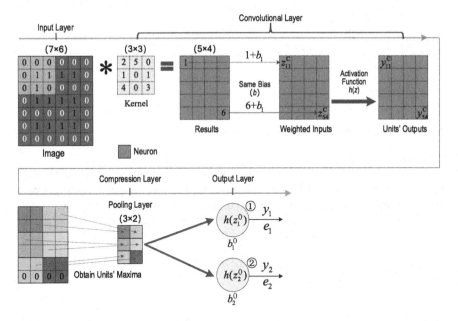

Fig. 3.41 The working process of CNN

(3) **The working process of CNN**

The working process of CNN also starts from the input layer, goes through the processing of each hidden layer, and finally outputs the results in the output layer. Its working process is similar to ANN's, but it adds a lot of processing processes to the image (it can also be understood as a preprocessing process). Here is a brief illustration of its working process using the images of the numbers "5" and "7" in Fig. 3.35 as an example, as shown in Fig. 3.41.

1. **Convolution calculation**

Taking image "5" as an example, the input data is a 7×6 matrix, and the input layer is set into 42 elements. The function of the convolution layer is to extract the contour features of the image. Let the convolution kernel be a 3×3 matrix. Each element is arbitrarily given, then this tiny matrix carries out the convolution operation on the original matrix.

After convolution calculation, the original 7×6 matrix is reduced to a 5×4 matrix, and the number of elements is changed from 42 to 20, which is significantly reduced (but the shape information of the image is still retained). The next operation is on the convolution result (5×4 matrix).

2. **Weighted input (sum)**

Like ANN, multi-layer networks also have a process of weighted input (summation) and activation. Let's first look at the weighted sum problem.

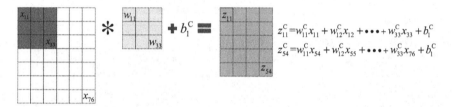

Fig. 3.42 General representation of weighted summation

Figure 3.41 shows that the weighted input has 20 elements since the convolution result is a 5×4 matrix. Add the same offset (b_1^C) to them respectively to get each weighted input. The expressions for two of these elements are given below:

$$z_{11}^C = [2 \times 0) + (5 \times 0) + (0 \times 0) + (1 \times 0) + (0 \times 1)$$
$$+ (1 \times 1) + (4 \times 0) + (0 \times 1) + (3 \times 0)] + b_1^C$$
$$= 1 + b_1^C$$
$$\ldots\ldots$$
$$z_{54}^C = [2 \times 0) + (5 \times 1) + (0 \times 0) + (1 \times 1) + (0 \times 1)$$
$$+ (1 \times 0) + (4 \times 0) + (0 \times 0) + (3 \times 0)] + b_1^C$$
$$= 6 + b_1^C \tag{3.35}$$

In the formula, the superscript "C" represents the convolutional layer. If representing the two above formulas using symbols, it can be expressed in the form shown in Fig. 3.42.

According to the expression in Fig. 3.42, it can be seen that elements of the convolution kernel ($w_{11}^C, \ldots w_{33}^C$) are the weights of the convolution layer. This explains why the value of the convolution kernel can be set arbitrarily.

3. **Activation (nonlinear transformation)**

By substituting each weighted input into the activation function, the output of each element can be obtained. The activation function can be selected according to the specific situation, not necessarily the Sigmoid function. The outputs of z_{11}^C and z_{54}^C are as follows.

$$y_{11}^C = h(z_{11}^C), \; y_{54}^C = h(z_{54}^C) \tag{3.36}$$

At this point, the job of the convolutional layer is done. Next, the 20-output data image needs to be further simplified to reduce the number of elements.

4. **Information compression (pooling)**

After the convolution processing of the image, the amount of data does not reduce too much, which needs to be compressed (a dimension reduction operation). The

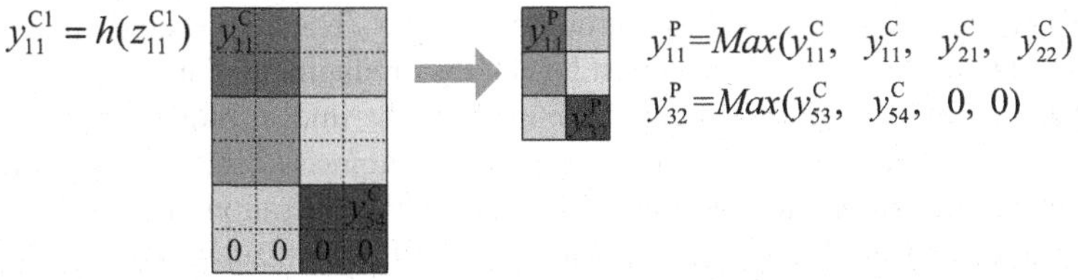

$$y_{11}^{C1} = h(z_{11}^{C1})$$

$$y_{11}^{P} = Max(y_{11}^{C}, \ y_{11}^{C}, \ y_{21}^{C}, \ y_{22}^{C})$$

$$y_{32}^{P} = Max(y_{53}^{C}, \ y_{54}^{C}, \ 0, \ 0)$$

Fig. 3.43 Compression method of image information

compression method is straightforward. The elements of the convolutional layer are divided into non-overlapping 2×2 matrices (use 0 to fill the spots without numbers). Then, the maximum value of each 2×2 matrix element is taken to form the compression layer element. This compressed layer is also called the pooling layer or sampling layer, as shown in Fig. 3.43.

After compression processing, there are only six elements, which is significantly reduced compared to the original 42 elements, and the amount of calculation is also reduced.

5. **Output results**

The compressibility layer consists of 6 elements, expressed in the form as shown in Fig. 3.44. The processing method is the same as the full connection processing method of ANN, as shown in Fig. 3.44.

So that's how CNN works. The basic idea of CNN is to conduct information processing in layers and stages. In practical application, one can set up multiple convolution kernels for processing according to the specific situation (the same convolution kernels slide on the image to extract the same feature. Multiple convolution kernels should be used if multiple features are extracted). After the extraction, multiple convolution layers will be formed. Thus, the amount of compression layer will also increase. But no matter how many more compression layers, the output layer must be calculated according to the actual connection.

CNN has several features. First, the convolution check is used to preprocess the output information to extract the contour features of the image. The convolution

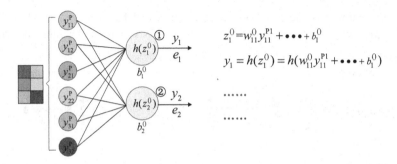

$$z_1^0 = w_{11}^0 y_{11}^{P1} + \bullet \bullet \bullet + b_1^0$$

$$y_1 = h(z_1^0) = h(w_{11}^0 y_{11}^{P1} + \bullet \bullet \bullet + b_1^0)$$

Fig. 3.44 Processing of the output layer

kernel's elements are regarded as a group of weights. They are shared (such as the nine weights in the above convolution layer), thus reducing the amount of calculation. Second, the resampling method compresses the image, reducing the number of elements and the amount of calculation. The compressed image is significantly reduced, but the original features are still retained. The operation mimics the human visual nerve, so even if a picture is compressed to a small size, it doesn't affect our ability to recognize it.

(4) **Weight adjustment—error backpropagation**

It can be seen from the working process of CNN that the convolution kernel is composed of a set of weights, and there is another set of weights between the compression layer and the output layer. The adjustment method is the same as that of ANN. The gradient descent method is still the principal, and the error backpropagation method is used to carry out a specific operation. Due to space limitations, this book will not explain the weight adjustment in detail. Interested readers may refer to the materials on deep learning.

(5) **Other multi-layer network algorithms**

Deep learning is a general term for multi-layer neural network algorithms. In addition to CNN, recurrent neural network (RNN) and generative adversarial network (GAN) are also commonly used.

In general, the multi-layer neural network is characterized by multiple hidden layers, linear and nonlinear operations, the output layer for summation and activation, and output results (Fig. 3.45).

Deep learning, which requires the support of large amounts of training data, becomes popular with the rise of so-called "big data" and is also closely related to improving the computing power of computers.

> The purpose of learning an algorithm is to understand the principle and implementation process, not limited to the application. Learn how to improve your algorithms to improve performance. Many toolkits are very convenient, but they block our understanding of algorithms. Therefore, do not rely too much on the toolbox to learn to write code based on understanding the principle. Through such a process, we believe that understanding various algorithms will reach a new height.

Deep learning is the most popular algorithm and has achieved good results in image recognition and speech recognition. CNN plays a leading and vital

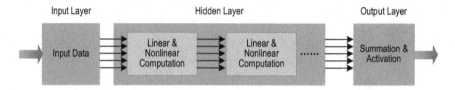

Fig. 3.45 The working process of a multi-layer neural network

role in developing AI. Deep learning is mainly used for classification problems, among which computer vision (essentially image recognition) is the most important. Suppose the intelligent robot is aware of the surrounding things. In that case, it will be beneficial (to identify materials and detect structural cracks, etc.) in the engineering field. The critical technology is image recognition. The input data (image information) and feature extraction are generally extensive for this problem, and ANN cannot do it.

3.5 Other Algorithms That Can't Be Ignored

There are many machine learning algorithms. ANN and CNN are just two of these categories, and we cannot ignore other algorithms. Although the mechanism of these algorithms is not based on the physiology of the human brain and is not completely intelligent, many algorithms related to statistics can solve some practical problems. This is sufficient. Maybe there will be better intelligent algorithms in the future, but it's hard to predict.

Generally speaking, any technology that uses a computer to train the data to obtain the hidden laws in the data and thus create the computer model is machine learning. Mathematics is the key to this kind of approach. The following is to introduce some algorithms according to the different learning methods.

(1) Supervised learning algorithm

Supervised learning is the primary way of machine learning, including many algorithms. The input data mainly determines or predicts a specific feature of output data (regression or classification). In addition to the ANN introduced above, linear regression, logistic regression, decision tree, Naive Bayes, K-nearest neighbor, random forest, support vector machine, and so on. These methods mentioned above are all supervised learning algorithms, which are all based on statistics.

The common characteristics of supervised learning algorithms are that they need a large amount of known data for training, have a clear goal, and know the results they want. The following is a brief introduction to K-Nearest Neighbor and Support Vector Machines.

1. K-Near Neighbor

K-Nearest Neighbors (KNN) is the most straightforward classification algorithm suitable for multiple classification problems. For example, a circle and a square represent two types of known sample points, and a sample point with an unknown attribute (represented by a triangle) needs to be categorized, as shown in Fig. 3.46.

The core idea of KNN is that "One takes the behavior of one's company." According to the distance between the known samples and the samples to be classified, K neighbor known sample points for each sample point to be classified are identified., This sample point is classified as having more known sample points among these K neighbor points.

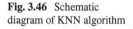

Fig. 3.46 Schematic
diagram of KNN algorithm

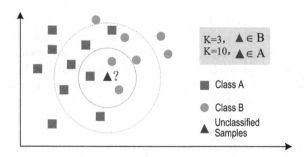

For example, K = 3, with the triangle point as the center to be classified, draw
the three nearest neighbor points according to the distance. Since the two neighbor
points belong to class B, the triangle points belong to class B according to the above
principle. If K = 10, then 6 out of ten neighbor points are class A; thus, the triangle
points belong to class A.

The classification of KNN is regular, which is different from the previous algorithm. The most significant advantage is that it is simple and easy to operate. It
supports multiple classifications and does not require training. It can be classified
directly according to data. However, the disadvantages are also significant. Different
classification results may be obtained when different neighbors are selected (i.e.,
different K values). How to overcome this shortcoming is still unclear.

2. **Support Vector Machines**

Before deep learning became popular, the hottest algorithm was statistics-based
support vector machines (SVM). This computing technology emerged in 1995, with
its strict mathematical theory and excellent performance, which became the leading
machine learning algorithm and set off the upsurge of statistical learning around
2000. Now, the classification of two types of feature points (class A and class B)
on a two-dimensional plane is taken as an example to explain briefly, as shown in
Fig. 3.47.

The horizontal and vertical coordinates in Fig. 3.47 represent the two features
of class A and class B, and the sample point (x, y) is regarded as two-dimensional
vectors. If the feature of the sample point is multi-dimensional, it will constitute a
multi-dimensional feature space. How to extract features is critical, which is the key
to determining the classification quality. Here is a further illustration using Fig. 3.47.

The line that distinguishes class A from class B is called the classification line,
also known as the classifier, and the line $ax + by + c = 0$ in the figure is a classifier.
The sum of the distances from the nearest sample point to the classification line
is called the classification interval in classes A and B. Its magnitude reflects the
reliability of the classification. The larger the interval, the higher the reliability of
the classification (Fig. 3.48). The most significant classification interval classifier is
the support vector machine, while the sample points closest to the classification line
are support vectors. Points A1, B1, and B2 in Fig. 3.47 are support vectors.

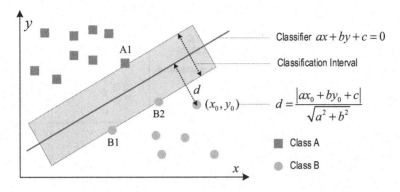

Fig. 3.47 Schematic diagram of SVM algorithm

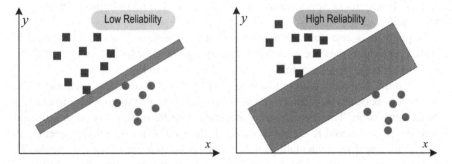

Fig. 3.48 Schematic diagram of reliability level of classification

For two-dimensional data, there are many classification lines. The goal of SVM is to find the line with the most significant classification interval (support vector machine). The essence of SVM is determining the coefficients a, b, and b to maximize the classification interval. For 3D data, a classifier is a plane. The classifier is a "hyperplane" for multidimensional data with more undetermined coefficients. How to determine these coefficients is the key. We still need to use known data for training, a complete set of mathematical theories for reference (many formulas), and the reader can refer to the relevant information.

(2) **Unsupervised learning algorithm**

Unsupervised learning algorithms are still under development, and only a few kinds exist. The common ones are K-means clustering, principal component analysis, Gaussian mixture model, etc. Its typical feature is to automatically find the data's potential regularity and effectively summarize it, which is also a statistical method in reality. The typical algorithm is K-means clustering (1956), the most important unsupervised learning algorithm.

Clustering automatically divides data into several groups to ensure that the same data group has similar characteristics. This method is derived from statistics and is

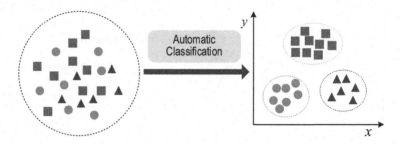

Fig. 3.49 K-means clustering

commonly used in data mining and analysis. As an unsupervised learning method, the clustering algorithm does not need training data, so it is an efficient data analysis method. It has crucial applications in many fields such as finance, medical care, and data mining.

Suppose it is two-dimensional data (with two features), and the difference is evident as long as they are placed in two-dimensional coordinates. In that case, the classification can be seen, which is intuitive, as shown in Fig. 3.49.

When the data has multiple features, it will not be visualized. In addition, if the differences between the data are not very obvious, it is also not possible to classify them with vision. Therefore, a set of general scientific methods is needed to deal with this problem, and K-means clustering is the method to solve this problem.

The idea and implementation of K-means clustering are relatively simple. To visually illustrate the classification process, we still take two-dimensional data as an example in Figs. 3.50.

① Determine the number of data grouping (classification) K. There is no unified regulation on determining K, and it can only be determined according to the actual situation. Let's make K equal to 2, divided into two categories.

② Determine the initial clustering center of each group (class) (there are 2 in total). You can take the average horizontal and vertical coordinates of all the data in each group as the coordinates of these two centers.

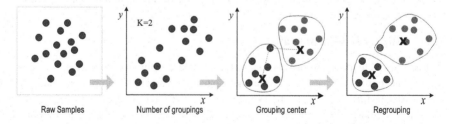

Fig. 3.50 Iterative calculation process of clustering

③ Calculate the distance between each point and the two clustering centers. According to the principle, the closer the sample point is to the center of a specific group, the more likely it is to belong to it. Thus, two groups were redivided.

④ Calculate the new center of all points in each class (the mean value of coordinates of all points), and repeat Step 3) until the division no longer changes.

Clustering is a method of automatically classifying data that does not rely on human knowledge, but what the resulting classification represents is unknown (see Fig. 3.8), and thus it is not intelligent. However, this approach has many practical applications, of which finding data outliers is the most important.

For example, through the clustering analysis of user information and consumption behavior in the bank card transactions process, we can find the abnormal users and determine whether there is fraud. In medical diagnosis, abnormal points can help diagnose the condition. It is also imperative to determine abnormal data for engineering construction, such as erroneous data, false data, etc.

(3) Reinforcement learning algorithm

Reinforcement learning, known as the representative of "behaviorism," is a new idea of computer simulation of the human brain. It mainly deals with the "process of action" issues and falls under decision or control. This problem concerns whether the course of action can bring the most significant revenue. It requires trial and error among all possible courses of action (strategy) to find the best to obtain the highest revenue.

1. Basic concepts

Reinforcement learning is developed based on the Markov decision process (MDP). Its basic idea is straightforward. If taking a course of action offers a better payoff, then further "reinforce" that course to achieve a better payoff; otherwise, change course. The best course of action will eventually come through repeated exploration. This is a class of algorithms that constantly learn and correct mistakes as they go along.

Reinforcement learning provides us with an interactive machine learning framework. This approach can be applied to any process decision and control problem. In conjunction with Fig. 3.51, here are some basic concepts, most of which derive from the Markov decision process.

Fig. 3.51 The basic ideas of reinforcement learning

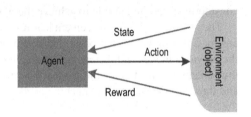

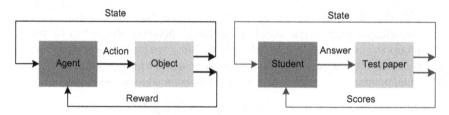

Fig. 3.52 Systematic description of reinforcement learning

In reinforcement learning, the entity making the action is called the agent. Robots, unmanned cars, intelligent road rollers, and AlphaGo are all agents.

The things outside the agent and associated with it are called the environment. The environment is the object of the agent's action. The environment has a feedback effect on the agent (the interaction between two systems), which is the main characteristic of reinforcement learning. For example, when the machine plays chess, the computer is the agent, and chess is the environment (object). They exchange information all the time.

The effect of an agent on the environment is called an action. Actions can change the current state of the environment (object) and can be observed (perceived) by the agent, and the environment rewards the agent called a revenue (also known as a reward). Revenue is the evaluation of the action and is a specific number. The bigger the revenue is, the better the action is.

It might be more intuitive to describe it in system language. The input (action) of the agent (system) to the environment will get two feedback information-the state of the environment and the size of the revenue response, as shown in Fig. 3.52.

For the agent, these two feedbacks directly affect the following action. Reinforcement learning guides the agent to make the most favorable actions according to the two feedback information to maximize revenues. How to do this involves the problem of strategy (scheme) optimization.

For example, students work on exam problems (Fig. 3.52). The student is the agent, the test paper is the environment (object), and the solution is the action. The answering situation is the state feedback and the revenue score obtained. You get all the answers correct and 100 points (the highest reward). If you get it all wrong, you get 0 points. Students adjust their learning strategies and constantly improve their ability to master knowledge through stimulation and punishment. In addition, it is not good to only look at past exam results and should have a long-term vision by considering overall planning to achieve the maximum revenue.

The inspiration for reinforcement learning comes from this human learning mechanism. Chess playing is a process decision problem to make the next move based on the previous moves and the change of the game (the opponent also drops the piece, thus changing the state of the game). The goal is to obtain the maximum revenue and thus win the game. The machine needs to learn various chess scenarios to find the best one to maximize revenue.

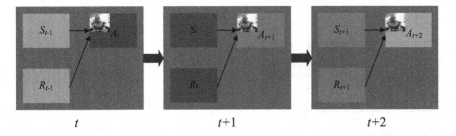

Fig. 3.53 Iterative decision-making process

So that's the basic concept of reinforcement learning. For a process decision or control problem, the agent can learn to choose what kind of action in what situation to get the best revenue through constant adjustment of actions and trial and error. There are various algorithms to realize this wish. How to choose needs to be analyzed in detail.

Because reinforcement learning algorithm involves more mathematical knowledge, such as dynamic programming, Markov decision process, Monte Carlo method, etc. A few notations (almost from MDP) must be introduced to ease expression.

As shown in Fig. 3.53, A_t is adopted for a dynamic problem representing the action of the agent at time t, which is determined by the state and income at time (t − 1); S_t represents the observed state of the agent at time t; R_t represents the revenue (reward value) that the environment feedback to the agent at time t.

At time t, action A_t produces a state S_t and a revenue R_t. They together determine action A_{t+1} at time $(t + 1)$, A_{t+1} produces S_{t+1} and R_{t+1}. This is an iterative process. It can be seen that in the process of iterative decision-making, revenue and state jointly control the action mode of the agent at the next moment.

Reinforcement learning aims to find the best action strategy for maximum revenue. The algorithm decides the way to find the best action strategy.

2. Q learning algorithm

Reinforcement learning has a variety of algorithms involving statistics, optimization theory, random processes, and other mathematical knowledge. It is also inspired by psychology and neuroscience. In conclusion, according to whether a model can represent the environment, algorithms are divided into model-based and model-free algorithms.

In model-based algorithms, the environment is represented by an algorithm. Given an input (action) to the environment, the algorithm predicts the state and revenue, and its action plan is controllable. This problem is relatively easy to solve and can be solved by dynamic programming techniques.

Because an algorithm cannot represent the environment in model-free algorithms, it is impossible to predict the resulting state and revenues directly after sending an action to the environment. It can only do trial and error to make the following action. For example, in chess, the player (agent) does not know what will happen to the board

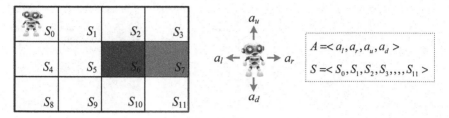

Fig. 3.54 Robot grid game

(environment) when s/he moves a piece, and it is difficult to determine whether the piece (action) is good or bad because s/he does not know what the opponent will do.

Reinforcement learning mainly refers to model-free algorithms, which is also the new trend in AI. Q learning is a model-free algorithm (developed in 1989) which is taken as an example for brief illustration.

Q learning is an algorithm that selects actions based on value functions. This is the most effective and universal reinforcement learning algorithm at present. Figure 3.54 shows a simple example of a robot grid game.

In Fig. 3.54, the robot's task is to go to the green grid S_7 from the initial position S_0 and learn the best way to walk. Here the agent is the robot, the environment is the grid, and the walking plan is the strategy. There are four possible directions for the robot to walk: move up a_u, down a_d, left a_l, and right a_r. Each cell has one state, and there are 12 states in total.

Use the revenue to judge the robot's walking (action). The revenue for walking to the green grid (S_7) is 1, and for the red cell (S_6) is -1, the revenue for every other cell is 0. As for the setting of revenue, there is no unified regulation. It mainly depends on the preferences. Subjective is a crucial feature of reinforcement learning.

The robot has four optional actions: a_l, a_u, a_r, a_d. The task is from S_0 to S_7. There are four routes, and the best route is $S_0 \rightarrow S_1 \rightarrow S_2 \rightarrow S_3 \rightarrow S_7$, but the robot doesn't know the best route, and the robot needs to be trained to find the best route, which is also known as the Markov decision process. The following is the solution to Q learning.

First of all, we need to define a value function Q = (S, A), which represents the expected value of the revenue that can be obtained by taking action A under the state S. The main idea of the algorithm is to combine state S with action A to build a table to store the value of Q and the form is shown in Table 3.2. The actions of the robot are guided according to the Q table.

The robot can choose its path according to the Q value if the Q table is known. For example, if the robot is at S_0, the possible actions include a_l, a_u, a_r, and a_d. If $Q(S_0, a_r)$ is the maximum, then select a_r which represents moving to the right to get to S_1; then continue to select the action with the maximum Q value ..., until it reaches S_7.

If the value of Q is unknown, training (learning) is required. The initial value can be set randomly, called an estimate, and the purpose of Q learning is to update

Table 3.2 Table of Q values

Q values	a_l	a_u	a_r	a_d
S_0	$Q(S_0, a_l)$	$Q(S_0, a_u)$	$Q(S_0, a_r)$	$Q(S_0, a_d)$
S_1	$Q(S_1, a_l)$	$Q(S_1, a_u)$	$Q(S_1, a_r)$	$Q(S_1, a_d)$
S_2	$Q(S_2, a_l)$	$Q(S_2, a_u)$	$Q(S_2, a_r)$	$Q(S_2, a_d)$
......	......	......	......	......
S_{11}	$Q(S_{11}, a_l)$	$Q(S_{11}, a_u)$	$Q(S_{11}, a_r)$	$Q(S_{11}, a_d)$

this table to an accurate Q table to guide actions. The updating formula of Q value estimate is carried out as follows:

$$Q(S_t, A_t) \leftarrow Q(S_t, A_t) + \alpha[R(S_{t+1}) + \gamma \max_a Q(S_{t+1}, A_{t+1}) - Q(S_t, A_t)]$$

(3.37a)

$$Q(S_0, a_r) \leftarrow Q(S_0, a_r) + \alpha[R(S_1) + \gamma \max_a Q(S_1, a_r) - Q(S_0, a_r)] \quad (3.37b)$$

Equation (3.37a) is the general formula; Eq. (3.37b) is the updated formula when the robot's initial position is S_0, and the action is a_r. where $R(S_1)$ is the revenue of the robot at S_1; γ is the attenuation rate, between 0 and 1. $\max_a Q(S_1, A_1)$ is the maximum value of $Q(S_1, A_1)$ at S_1. The action is a_r. α is the learning efficiency; $R(S_1) + \gamma \max_a Q(S_1, A_1)$ is the target value of $Q(S_0, a_r)$. The difference between the target value and the deviation is the error. See Fig. 3.55 for the above process.

Q learning algorithm is straightforward. It performs a different action in each state to observe the revenue iteratively. This method is based on the condition that the state set is known. If the unknown state occurs, it is impossible to predict the unknown state. In addition, the Q table can also be trained by ANN.

In addition to the algorithm based on value function to select actions, reinforcement learning also has an algorithm based on probability. Actions are selected according to the probability, limited by space, and will not be presented here.

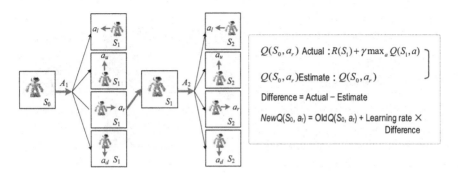

Fig. 3.55 Update process of Q value

Reinforcement learning is a type of algorithm related to dynamic decision and control. The reinforcement learning algorithm can be used in decision-making, control, finding the best solution, and other problems.

In addition, attention should be paid to the multi-agent theory, which began in the 1990s. When multiple agents work together, reinforcement learning can help improve incomplete perception problems, complex computational problems, reward allocation problems, etc.

When solving a complex problem, various algorithms are generally required to be used comprehensively, such as the famous AlphaGo, a combination of deep learning and reinforcement learning. Therefore, in practical applications, it is not necessary to be limited to a specific algorithm but to learn to choose it flexibly according to the specific situation.

3.6 Return to Reason: Look at AI Objectively

After looking at the principles of machine learning algorithms, let's focus on the rational, dispassionate, and objective view of AI.

Since its emergence, AI has experienced three stages of development: initial logical reasoning, knowledge-based expert system, and the current machine learning. In addition to statistical learning and genetic algorithm, ANN-based deep learning is a new stage of development, as shown in Fig. 3.56. Objectively speaking, today's AI is equivalent to machine learning (which is why this chapter is titled machine learning). How AI will evolve in the future is still an open question.

The scope of AI is not clear at this moment. Many data processing methods with no intelligence in statistics are included in machine learning. Such algorithms (linear regression, decision trees, Bayes, etc.) are far from what AI requires. In addition, a variety of bionic algorithms, fuzzy mathematics, cellular automata, game, and other methods have been included, but these technologies are also different from AI.

Deep learning is the most mysterious part of machine learning, which is not mysterious but a formal imitation of the brain's physiology. In essence, convolution

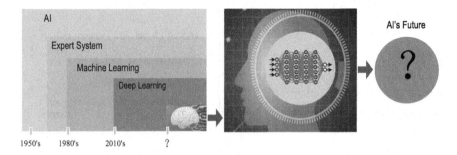

Fig. 3.56 Where AI is heading to

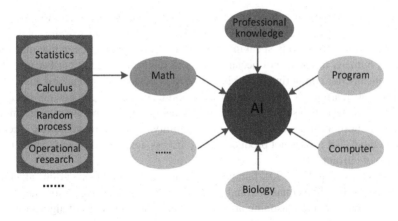

Fig. 3.57 Various knowledge underpinning AI

calculation, weighted summation, activation operation, and resampling (pooling) are mathematical operations of different forms that have nothing to do with the physiological structure of the brain. Reinforcement learning is the classic Markov decision process (MDP).

The current AI with machine learning as its core is still in mathematics in its algorithmic nature. Mathematics has naturally become the foundation of the current AI. The knowledge from other subjects is also indispensable, as shown in Fig. 3.57.

Among all the knowledge supporting AI, besides mathematics as the foundation, biology only serves as inspiration and guidance, computers and programs are tools for implementation, and professional knowledge (such as roads and railways) is essential for applying machine learning in different professional fields.

When we master the machine learning algorithm, the decisive role of all knowledge behind it will be clear. At this point, do you still believe that AI built from this knowledge will produce human-like intelligence? Still worried about AI posing a threat to humanity?

Since the current AI does not have actual human intelligence, how can the emergence of various intelligent phenomena explain? Such as chatting robots, a human cannot defeat AlphaGo, and so on.

Most of the current AI is driven by data. Algorithms are used to learn the known data and make predictions for unknown situations. Machines work well as long as they are known to cover a wide range of data. Generally speaking, it lets the machine learn as much data beforehand to adapt to various situations. With this in mind, it is easy to understand all kinds of intelligent phenomena.

This is how chatbots—an essential goal of AI research—enable machines to communicate verbally—work. When humans chat with machines, they sometimes think they are talking to humans, not programs, and feel competent. However, if the answer is not within its predefined range, the conversation will not be correct (perhaps nonsense, depending on how the programmer designed it).

AlphaGo trained for a long time to defeat a Go master, but later AlphaGozero took only 40 days to defeat AlphaGo, in which reinforcement learning played an important role. Objectively speaking, it is an inevitable phenomenon that machines surpass humans in chess, not because they are really intelligent, but because they have much better computing power based on the rules of chess than humans.

Machine learning is designing an algorithmic model to process the data and produce our desired results. We can continuously optimize the algorithmic model to form a more accurate data processing capability. It is a "new numerical calculation method," but it does not produce consciousness, let alone thinking, which is the essence of current AI!

For example, in the 1990s, famous mathematics could deduce various algebraic equations, differential equations, and symbolic relations and solve various analytical problems. Its machine reasoning ability is better than many current algorithms, but no one said this software is intelligent.

Does it make sense to develop a technology that lacks true human intelligence? The answer, of course, is yes. AI can use computers or computer-controlled devices to simulate human work, such as perception, learning, analysis, reasoning, judgment, communication, and decision-making. Although it is auxiliary, it is pretty good, and there is no need to wonder whether it has the intelligence of human beings, especially in the field of engineering.

There are specific "rules" to do things. Machines will do them better than humans. The most critical applications are complex calculations and repetitive mental tasks. But remember, machines don't create. They don't think. Human does.

There are also concerns about whether AI has safety and ethics issues. This is a software (program) error problem. If something goes wrong with the intelligent algorithm when programmed, something unexpected could happen, like the possibility that intelligent robots could attack humans. If it's intentional, it's the equivalent of a computer virus program.

3.7   Machine Learning in Engineering Projects

Machine learning is the core of current AI, replacing part of human mental work. At present, intelligent construction is still a new thing. As a tool to deal with problems, effectively applying machine learning is the key. As the Chinese saying, "stones from other mountains can rival jade," we can learn and draw lessons from the successful experience of AI applications in other fields to inspire us to apply AI to engineering projects.

Let's take a humanoid robot (mostly visualized) as an example to see what it already has and how it relates to engineering.

Figure 3.58 shows that humans mainly understand the external world through sensory organs. In the case of machines, these functions are achieved through various sensors. After learning and mastering specific skills, the machine will comprehensively analyze the perceived information and decide what action to take. If you think

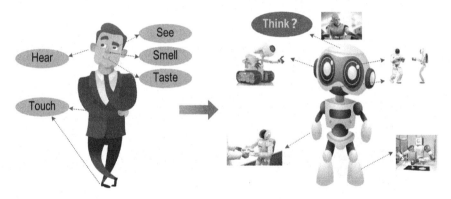

Fig. 3.58 Human versus machine

about human behavior in engineering projects, it's the same process. In this case, we can try to make machines assist humans in doing some work in engineering construction, which is also the essential requirement of intelligent construction.

The scope of work that machine learning can do in engineering projects can refer to some applications of AI in other fields. Currently, the technologies with potential application prospects in engineering projects include intelligent control, data mining, machine vision, fingerprint recognition, face recognition, intelligent search, intelligent service, intelligent management, etc. Behind these technologies are the applications of various algorithms. For example, computer vision is carried out by deep learning.

Understanding these applications will help us flexibly apply these technologies in engineering projects. Combining these machine learning algorithms with engineering projects is the work that needs to be carried out in intelligent construction, which sometimes requires skill and creativity. For example, deep learning can be applied to recognize topographic and geomorphic features in the design phase and damage images in the maintenance phase. In the construction phase, intelligent control technology can be applied. Management can be linked to fingerprint recognition, face recognition, intelligent services, etc. All the mental work involving human beings can be processed with appropriate machine learning methods to assist us in our work.

For inspiration, Fig. 3.59 shows the structure diagram of two ANN applications. We can apply these ideas creatively in engineering projects according to the problems we face.

In Fig. 3.59, the left figure contains two ANN networks. The secondary network has some influence on the primary network, so the output of the secondary network is passed to the primary network. This method can deal with major and secondary influencing factors, such as road damage, which has primary and secondary factors.

Figure 3.59 shows the network structure combined with ANN and wavelet analysis. The wavelet basis function is the activation function (transfer function) of the hidden layer, which can be used for fault diagnosis, etc. Of course, it can also

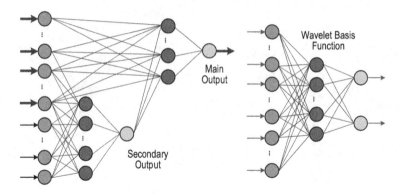

Fig. 3.59 Flexible application of ANN

be combined with other methods, which must be selected according to the actual problem.

In particular, the application of reinforcement learning in engineering projects should be emphasized. Reinforcement learning can be used for project management and service work. It can also be used for dynamic control of construction machinery, such as the cooperative work of multiple construction machines.

In addition, in engineering projects, no matter in the design, construction, or maintenance phases, there is a large amount of knowledge and experience accumulated and summarized over a long time. There is certainty and uncertainty, and it may be fuzzy. The "expert system + machine learning + other technologies" approach can be adopted for this problem. Such as using the neural expert system the rules with trained ANN to represent so that you can learn; For uncertain knowledge, the probabilistic expert system can be used. For fuzzy knowledge, the fuzzy expert system is used. In short, all algorithms can be used in combination without too much constraint.

Machine learning mainly solves classification and prediction (regression) problems. This requires us to generalize the problems in the project and see what kind of problems they belong to. In the past, most machine learning applications in engineering have been predictive problems—predicting future behavior based on a target's past behavior. But the classification problem is perhaps more critical and needs to be studied carefully, as the essence of intelligent design is the classification problem. Choosing appropriate tools according to specific problems is still relatively complex. We need to accumulate more experience in practice.

There is no doubt that machine learning will have more and more applications in engineering projects. Still, as far as the current technology is concerned, it needs the support of a large amount of existing data, and data will play a mainstay role in intelligent construction.

Chapter 4
Entering a Data Era

Abstract This chapter mainly introduces the primary content of data science. It includes the generation and essential characteristics of massive data, the difference between big and small data, causality and correlation, the problem of knowledge extraction in data, data mining, and cloud computing. The basic concepts of edge computing are briefly explained. Finally, data security is discussed.

4.1 Embracing the Data Era

Data is everywhere in our daily life. Whether the information is sensed by a human body or detected by an instrument, it can all be regarded as data. In engineering construction, a wide variety of data such as measurement, design, experiment, and management data have formed the "language of engineers," whose importance is self-evident. With the development of information technology, all data can be recorded and stored by computers. The Internet and Internet of Things enabled the possibility of "all things digital." Since then, data has been extending from professional fields into people's daily lives (Fig. 4.1). The time for data-driven technology has come.

Industries have benefited significantly from seeking patterns and trends in the data. Data have been given unprecedented priority in almost every field, especially since the rise of the Big Data concept in 2010. Now "Let the data speak" has become a catchphrase for professionals and laypersons. Next, let's discuss the basics of data.

(1) **Digital, data, and information**

When it comes to data, we must start with the concept of digital. Digital City, Digital Economy, and similar terms are overwhelmingly used in today's world. It is hence worth firstly clarifying the concept of digital.

Digital is a professional term. Its full name, digital quantity, is proposed as relative to analog quantity in the context of electronic circuits. Analog electronics involves quantities with continuous values, i.e., analog quantities. Digital electronics involves quantities with discrete values, i.e., digital quantities. Digital quantities vary in discrete steps of predefined time duration while changing from one point to another,

© China Railway Publishing House Co., Ltd. 2023
G. Xu and D. Wang, *Introduction to Intelligent Construction Technology of Transportation Infrastructure*, Springer Tracts in Civil Engineering,
https://doi.org/10.1007/978-3-031-13433-3_4

Fig. 4.1 Data everywhere

with only "high" and "low" levels. Binary zeros and ones usually represent them. Different combinations of zeros and ones can have different meanings. For example, 1,011,000 in binary means decimal number 88.

Digitization is the process of converting an analog (decimal) quantity to a digital (binary) quantity. Digitalization is not a new concept that has grown with the development of computers. Digitalization is sometimes called computerization or informatization; they fundamentally have the same meaning but different terminologies. Making information digitalized can facilitate computer storage and processing capacity, significantly improving data management efficiency.

Data is a representation of facts. Any representation of facts using words, figures, speech, graphics, images, sound, and video can be regarded as data. In computer science, data is presented as binary zeros and ones. All the symbols input into a computer and processed by a computer program are called digital data, or data for short. Data era is the abbreviation of digital data era.

Data is different from numbers. A number is a symbol and can be a part of data. Numbers can be expressed in binary or decimal digits, which have no specific meaning by themselves. They will become meaningful after they are interpreted as data. For example, 60 contains the numbers (digits) 6 and 0, which can be used to describe a person's age or a test result.

Information is usually defined as knowledge communicated or received concerning a particular factor or circumstance or any pattern that influences the formation or transformation of other patterns. The relationship between data and information is an interconnected one. Data is raw facts such as phone numbers or addresses, the form, and the information carrier. Information is the organization of these raw facts in a meaningful manner. These two cannot be separated, and a data era always comes with an information era.

In many cases, data and information are interchangeably used. The digitization of information is converting analog data into computer-readable binary data; for the same information, its content remains unchanged.

(2) Massive Data production

Massive Data (fashionably called big data) is a collection of data that is huge in volume yet growing exponentially with time. Data with such a large size and complexity that none of the traditional data management tools can store or process it efficiently. The magnitude of Big Data can reach terabyte (TB) or petabyte (PB), and one of its primary sources is the Internet (Fig. 4.1). A considerable volume of data is produced with the popularity of social media networking, e-commerce, and mobile communications. The rising and growth of the Internet of Things (IoT) allow physical objects (or groups of such objects) embedded with sensors, processing ability, software, and other technologies to connect and exchange data with other devices and systems over the Internet and other communications networks.

IoT and the Internet form the primary production source of today's Massive Data. They connect science, technology, business, engineering, banking, healthcare, and almost all fields to produce, exchange, utilize, and manage data.

(3) The value of data

The spread of computers and the Internet has turned all information into digital data, while the value of the data has genuinely attracted people's attention. For example, by analyzing people's online search patterns, Google successfully predicted the timing and location of the H1N1 swine flu outbreak in the United States in 2009, giving public health agencies valuable information for making strategies and decisions. Many application cases like this make the public realize the power and value of data and gain a strong interest in data. The rise of AI has raised the value of data to an even greater level.

Data are given different values and interpretations in different fields. Benefits in business, engineering quality, and medical field health conditions can all be quantified with data. Big Data not only has immense commercial value (such as revenue generation and cost-saving), but they also generate social, scientific, technical, and engineering values when the different analysis focuses are held. To sum up, the potential value of massive data is infinite, and we need to explore it.

On the other hand, accessing data is difficult (data ownership issue). Large Internet companies mostly own massive data, which can utilize the data resources, processing power, and statistical technologies under their control to make correlations and gain values from the data. However, it is not easy for ordinary users to obtain massive data. Therefore, data circulation and sharing is a big issue.

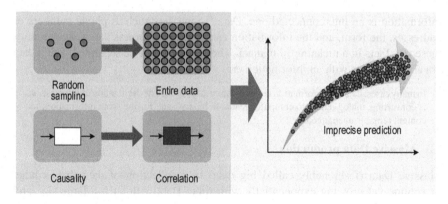

Fig. 4.2 Basic features of the data era

(4) **Basic characteristics of the data era**

In the data era, random sampling is no longer the primary strategy to deal with the massive data; instead, the whole data population is often of interest. The data analysis's purpose has also shifted from causation to correlation, focusing on the existence of trends rather than the accuracy of the relationship prediction. These essential characteristics of the new data era, as presented in Fig. 4.2, will profoundly impact the way data are processed and how we view the world.

In the past, due to the limitations of acquisition, storage, and computation conditions, the sufficiency of data is what people usually care about (in fact, never sufficient). With the rapid advancement in electronic and information technology, processing massive data is no longer a concern. Hence, data have become a commercial capital that can create substantial economic benefits. One can easily find thousands of examples of how the wise usage of data has saved costs and labor, extended the service life, and enhanced infrastructure safety.

In the future, people's lives will be increasingly dependent on data. No matter in which field, those who understand the importance of mass data and master the processing and utilization of data will excel, which is the same for intelligent construction.

4.2 Big Data Versus Small Data

Data of everything and everywhere has become the symbol of the data era, among which the most popular word should be Big Data. The concept of big data is not always clear to the public. It originated in astronomy and genetics and was put forward because the amount of data needed to be processed was too large for the storage and processing power of computers or conventional databases at the time. It has become a prevalent word after being amplified by social media. The author believes this word will fade out of people's attention one day, which is why this book uses the word massive data a lot. The opposite of big data is small data. Now let us discuss the relationship between big data and small data.

A vast amount of data quantity (e.g., in PB or TB levels) is the most significant feature for massive data, but it is not essential. What matters the most is whether the information contained in the data is comprehensive. For an event, if a batch of data contains all the information, it is all data; there is no need to increase the quantification of data. For example, if a $10^3 \times 10^3$ resolution is sufficient to describe the details of a portrait, increasing the resolution to $10^9 \times 10^9$ or even higher will not add additional value. The quality and quantity of data are often evaluated by "value density," a concept borrowed from the logistics industry to determine the best mode of transport for goods based on weight and value. Value density in big data is referred to the ratio of event relevance to the size of the data, i.e., how many facts per gigabyte. A relatively low-value density is one of the challenges of dealing with big data compared to the data we manage in traditional systems. In other words, big data is not simply a large quantity of data but should contain more comprehensive information.

From the statistics perspective, big data is equivalent to a population of events, while small data is a sample. The relationship between big and small data is associated with the amount of information in an event. The amount of data that can fully reflect the event is big data; otherwise, it is small. In other words, the size of data should be determined based on the amount of information needed to represent an event. In that context, it might be more appropriate to call big data "all data" (Fig. 4.3). Note that large amounts of data are not necessarily big data. For example, in the population survey, a census is the population and is big data. Although the amount of data is large, it is a sample and still considered small data if it is a spot check. Therefore, one should not distinguish the big and small data simply by the data sizes but by whether the data contains all the event facts.

On the other hand, if the data size is not enough, they often may not represent the entire event. Using the flu prediction example mentioned above, if the number of search records collected was not significant, they might not represent the general

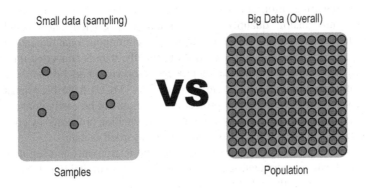

Fig. 4.3 Big data versus small data

N1H1 flu trend of the United States in 2009 for decision making. Although the data size was considered an issue, it is emphasized here whether it can contain comprehensive information on the flu prediction event. Therefore, comprehensiveness is the essence of big data. For social, e-commerce, health care, and Internet search issues, one needs enormous data to tell the whole story.

In many cases, we may not have access to the whole data set, or the cost of access is too high, and we will have to deal with small data (small samples). It is thus valuable yet challenging to use small data to find patterns in machine learning. All roadway and railway engineering tests are sampling tests and belong to small data. However, due to the variability of building materials, whether the sampled data is representative enough has always been controversial. If one can get data at every point in the structure, it is big data (all data), even if the data size is much less than a petabyte. Sensors embedded at every point in the structure can collect all data, but the cost is too high, and too many sensors may affect the structure's performance. One potential alternative is to use continuous detection techniques, as discussed later.

4.3 Digitization of Engineering Information

Engineering information generally refers to all the contents related to engineering construction. In the data era, information is presented by various forms of data. Data can be numbers, a combination of words, letters, numerical symbols, graphics, images, videos, or audio. For example, date, weather, construction records, and material properties are all data and part of engineering information.

The information encountered in engineering construction needs to be digitized for computer storage and processing. The digitization of engineering information falls into two categories: (1) the information directly entered into the computer. Such information, including text and data, is automatically converted into binary digits. (2) the data measured or collected by various instruments and devices. Such data are generally analog quantities and need to be digitized by particular devices.

The digitization process of engineering information follows three steps: Firstly, various engineering information is collected (perceived) by sensing devices and converted into analog quantity. Then sampling is carried out through A/D (analog/digital) conversion device to change the continuous quantity into discrete quantity, i.e., discretization and quantization. Finally, data is converted into binary code, i.e., digital data. The digitization process is shown in Fig. 4.4.

Currently, microprocessors or microcontrollers (MCU) have been used in all digital conversion devices so that project information can be easily converted into digital quantities and processed by computers. Information processed by computers, such as videos, images, texts, and sound, can be regarded as data.

In engineering construction, most experimental instruments and testing devices can automatically convert analog data into digital data so digital transformation can be realized conveniently and efficiently.

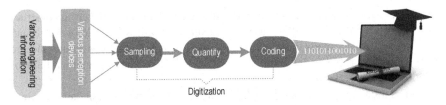

Fig. 4.4 Digitization process of engineering information

It is worth noting that any information (e.g., a word), when entered into the computer via a keyboard, is automatically converted into binary digits. When displayed on the screen, it is converted back to its original form (i.e., the word).

4.4 Causality and Correlation

Data analysis typically consists of two purposes: identifying causality and establishing correlation. Causality, also called cause and effect, is a deterministic relationship in nature. It is the influence by which one event, process, state, or object (a cause) contributes to the production of another event, process, state, or object (an effect) where the cause is partly responsible for the effect, and the effect is partly dependent on the cause. Causal analysis is an essential skill that people in science and technology (including social science) must have and a method that engineers must master. People always ask "why" to seek the causes and identify solutions to a problem. The answer to the problem is the "cause."

Figure 4.5 shows two examples of causality, although many causality examples may not be expressed analytically. The figure on the left shows a force F applied to an object M, causing an acceleration A, where F is the "cause," and A is the "effect." Newton's second law explains that force is the root cause of the motion of an object. The figure on the right shows a relationship between x and y, where $x = 3$ is the reason why $y = 45$.

The occurrence of an event may be related to multiple causes. Figure 4.5 shows a relatively simple case in that causality clearly describes the cause and effect of an event, and the relationship is deterministic and relatively easy to control. In system language, this is a "white box" problem (see Sect. 2.3), and everything is under control.

Although essential, a definite answer to causality is not always available. We often only know what factors contribute to the outcome but don't know why. For

Fig. 4.5 Example of causality

example, engineers can face various engineering problems in engineering construction and need to find out the causes based on their professional knowledge and experiences. But in many cases, they only know which factors contribute and are related to the problems but cannot specify why. This leads to another kind of relationship—correlation.

Correlation is the statistical interdependence, a phenomenon abundant in nature and human society. It describes the relationship between different variables, linear or nonlinear. Correlation analysis plays a significant role in statistics, which has always been a data analysis method for small data. In the era of big data, correlation analysis is based on massive data. It does not need to make various assumptions and has lower requirements on accuracy, which is essentially prediction analysis.

Many unexpected results can be obtained through correlation analysis, particularly in the sales market. According to the correlation analysis of many sales records, vendors found that placing two previously unrelated items together might increase sales, as shown in Fig. 4.6. One classic example is the "beer and diapers" story: young fathers would make a late-night run to the store to pick up Pampers and get some Bud Light while they were there. Putting egg tarts with a flashlight during a hurricane has also been beneficial for increasing sales. From the vendors' perspective, selling merchandise A and B is more critical than justifying why they should be placed together. Correlation analysis cares more about the results than the cause, while behavioral or social science studies will analyze the reasons for the correlation.

Anyone with an online shopping experience probably has recognized that they will continuously receive recommendations on relevant products as soon as they buy or search for specific products. This is realized by the online recommendation system, which recommends products, articles, and other information to users according to their purchase and browsing records based on correlation analysis. Relatively, correlation analysis has a lower requirement for accuracy.

Applying correlation analysis technology in the data era can be an exciting topic for engineering construction. It enables the integrating and synthesizing different types and data sources to establish meaningful correlations and achieve essential conclusions.

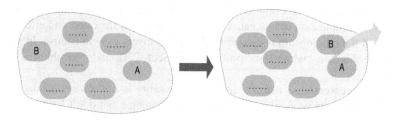

Fig. 4.6 Different product combinations can increase sales

4.5 Knowledge and Data

In addition to correlation analysis used mainly in business, data contains a wealth of knowledge. Knowledge comes from the awareness and experiences accumulated in practice, which can be expressed in various ways, such as natural language, mathematical language, physical models, words, and music. From the perspective of AI, knowledge is the information structure formed by connecting relevant information. Figure 4.7 shows that knowledge can be formed in two types of structures: rules (i.e., If–Then structure) or facts.

In artificial intelligence, for a machine to simulate some of the intelligent behavior of a human, it must know. The emergence of vast amounts of data has undoubtedly contributed to developing supervised learning methods. An expert system can also be designed to emulate the decision-making ability of a human expert and solve complex problems by reasoning through bodies of knowledge. For example, when combined with ANN, a neuro-expert system can be formed with the ability to learn.

In the past, people acquired knowledge mainly from practice and books; now, they rely more on vast amounts of data to gain knowledge. Various data processing methods, including machine learning, have been used to extract knowledge from massive data. Figure 4.8 shows the evolution from data to wisdom.

From data to information and then to knowledge, it is a process of qualitative change. Knowledge is contained in data, and information is the connotation of data. Once the knowledge is obtained from the data, it needs to be fed back into the machine to gain some wisdom and apply the knowledge. This process is more complex than producing intelligence itself.

Data is the language of engineers. In engineering, a common saying is "let the data speak for itself." The data here refers to the information and knowledge contained in the data, which is more meaningful than the data itself.

For example, when evaluating the compaction quality of roadway subgrade, if the sampling test obtained a 90% degree of compaction, the direct information to engineers is that the compaction quality of the testing spot achieved 90% of the maximum density and was not satisfactory (Readers can think about whether there is some wisdom in this?). However, engineers should be aware that this could imply that certain testing spots do not meet the compaction requirement, not necessarily the entire roadway section.

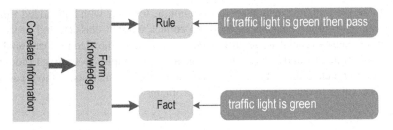

Fig. 4.7 The formation and expression of knowledge

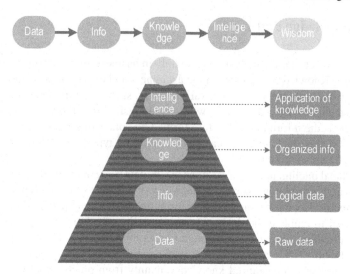

Fig. 4.8 From data to intelligence to wisdom

To fully understand the construction quality of the subgrade layer, other data such as degree of compaction for full roadway coverage, material properties, compaction equipment, and operation parameters should also be collected. From these data, the compaction quality information can be obtained, and the knowledge related to construction quality can be further extracted. Such knowledge of construction rules and findings can form theories and methodologies and become the basis for intelligent compaction.

Extracting knowledge from data (known as knowledge mining) has become a research focus in many industries. In intelligent construction, the knowledge-based expert system has attracted lots of attention. The extracted knowledge needs to be further organized, and the developing knowledge graph technology may also be used to provide new ideas and directions.

4.6   Data Analysis Methods

Data in the modern information society is a unique resource with high added values. Selecting appropriate analysis methods to collect, extract, classify, and process a variety of data is crucial to acquiring information and knowledge. Data analysis exists in any industry and covers a wide range from the simplest averaging to complex calculations (such as the fast Fourier transform FFT). The analysis methods can also differ according to data types and specific problems.

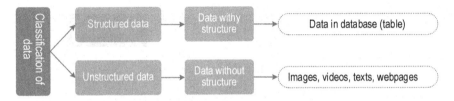

Fig. 4.9   Classification of data

(1)   **Data classification**

The first step of analyzing data is understanding what type it belongs to. Data can be classified in many ways, such as qualitative and quantitative data according to the nature of data or digital and analog data based on an expression. Another critical data classification is structured and unstructured data, whose characteristics are shown in Fig. 4.9.

Structured data conforms to a data model, has a well-defined structure, follows a consistent order, and can be easily accessed and used by a person or computer program. Structured data is usually stored in well-defined schemas such as Databases and is generally tabular with columns and rows that clearly define its attributes. Common examples of structured data include experimental data and construction reports. Although only about 15% of the total data population, structured data is widely used in engineering and other fields.

Unstructured data is not arranged according to a pre-set data model or schema and, therefore, cannot be stored in a traditional relational database. Text, multimedia, documents, email messages, videos, photos, audio files, and web pages are examples of unstructured content. Unstructured data is more common than structured data, and its volumes are increasing much faster than the rate of growth for the structured database. Although abundant in information, unstructured data has historically been challenging to analyze. With the help of AI and machine learning, new software and tools are emerging that can search and process large quantities of unstructured data to uncover valuable intelligence.

(2)   **Three types of ways to understand the world**

Data analysis is to find some regularity in a pile of "chaotic" data to rationalize the value of data and support decision-making.

The essence of data analysis is to understand the world. Although there are many data analysis methods, they can be summarized into three categories, corresponding to three cognitive approaches to learning, as illustrated in Fig. 4.10.

The modeling methods establish models based on scientific theories to analyze the attributes of an object, which belongs to the white box problem. These methods construct models based on scientific theories and do not need data in advance. For example, the mechanistic approach determines the displacement of the roadway base

Fig. 4.10 Three types of analysis methods

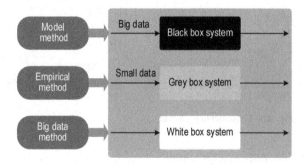

layer based on the elastic half-space theory. Once Young's modulus and Poisson's ratio of the base layer are known, the base displacement can be calculated according to the input stress level using a pre-derived equation.

The empirical method belongs to the grey box problem, with a known model structure but some unknown parameters. Traditional statistical regression and empirical formulas fall into this category. For example, a regression formula $y = ax + b$ can be used to describe a known linear relationship. Some data (small data) is required to determine the parameters (a, b), so the formula is established to determine output y based on any input value x.

The Big data method does not rely on specific models, but a lot of data belongs to the black-box approach. This method focuses on seeking patterns in data, such as the various machine learning algorithms.

(3) **Data visualization**

Data analysis aims to identify rules and patterns from a variety of data. In many cases, data presented in graphical form is much easier to comprehend and digest than a tabular presentation. The need for data visualization is exceptionally high when dealing with a large amount of data. Data can be visually presented using diagrams of correlation, trend, distribution, scatter, map, and many other formats and styles. It would be hard to imagine what can happen if a map is turned into a data table.

When dry data are turned into colorful graphics and images, their readability and attractiveness are much improved; more importantly, some information hidden in the data becomes more straightforward. Figure 4.11 shows some examples of turning data into graphical presentations.

Visual representation is essential to almost all analysis methods. Compared with traditional charts and tables, data visualization demonstrates information using more dynamic and interactive approaches. Interactive displays are often used in business and engineering to access and manipulate electronic files, illustrate critical conditions, and diagnose problems.

Data visualization is still an emerging concept whose technologies are under rapid development. Everyone should pay close attention to these developments to better develop the visualization technology of engineering data.

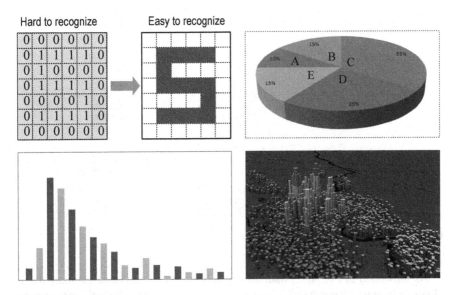

Fig. 4.11 From data to graphics

(4) **Statistical methods and machine learning methods**

Data analysis methods can be categorized into two main groups, statistical methods, and machine learning methods. Each group of methods covers a variety of specific methods.

Statistical methods have always played an essential role in data analysis. In addition to the traditional correlation and regression analysis, other analysis methods such as Principal Component Analysis (PCA) and Cluster Analysis have also been developed and widely used. Most of these methods have been integrated into data analysis tools such as Excel, SPSS, SAS, etc. With a basic knowledge of statistics, one can easily use these analytics tools to perform data analysis.

Machine learning methods, introduced in the previous section, are now gaining more popularity due to their close association with AI technology. Many statistical methods have been incorporated into machine learning, such as clustering, random forests, decision trees, Naive Bayes, etc. It is more important to understand these methods' basic principles and application conditions than to emphasize their specific classification.

(5) **Data mining**

Data mining is not an algorithm but a process of extracting and discovering hidden information in large data sets using various data analysis methods. The term "data mining" can be traced back to the 1990s in business and financial analysis. It is now used in many industries to extract potentially useful information from large quantities of data.

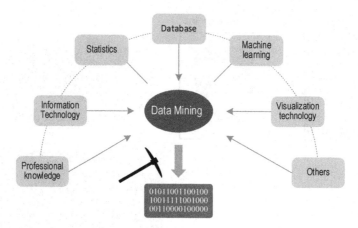

Fig. 4.12 Disciplines involved in data mining

As illustrated in Fig. 4.12, data mining relies on the knowledge of many disciplines, among which databases, statistics, and machine learning have the most significant impact. Databases provide data management techniques, while statistics and machine learning provide data analysis techniques. Data mining can be realized by specialized application tools and requires little statistical background.

Data mining is based on statistics and machine learning with a core of pattern recognition. It performs semi-automatic or automatic analysis using integrated tools and techniques and requires little human involvement.

Currently, most data mining techniques come from machine learning, such as cluster analysis, association analysis, and anomaly analysis.

(6) **Big data analysis**

Big data was initially proposed because of the computer's lack of storage and processing power, but this problem has been improved with computer performance enhancement. Therefore, for large quantities of data, the basis of its analysis methods is still statistics and machine learning. Data mining can be used to complete association rule analysis, cluster analysis, time series analysis, and outlier detection, which are critical to transforming data into intelligent resources.

Big data analysis consists of visualization analysis, data mining, predictive analysis, data quality, data management, etc. For massive data processing, most of these analysis methods have been integrated into some data analysis platforms, such as the Hadoop platform.

It should be noted that some new terms in the field of big data can essentially have similar meanings as the traditional concepts. For example, "data cleaning" is the previous data pre-processing. The term "data warehouse" (1990s) was also developed from the database concept; the "warehouse" has a larger capacity and can store data sets from different sources simultaneously.

(7) **Cloud computing and edge computing**

Cloud computing and edge computing have been gaining popularity in recent years. They are not considered algorithms but computing or service models that are particularly suitable for massive data. Suppose we want to do a massive data calculation. Still, the local computer's computing power is insufficient, and the remote computing center has a high-performance computing system. We can then request the computation task being done by the computing center. In the past, we need to go to the computing center to perform the computation in person; now, we can use the local computer to connect to the distant computing center through the Internet to complete the calculations. Figure 4.13 shows that cloud computing, traditionally called "online computing," is computation using network resources.

Cloud computing uses a network of remote servers hosted on the Internet, rather than a local server or a personal computer, to store, manage, and process data. High-performance computers usually realize centralized data processing through parallel computing. Instead, edge computing is about providing services close to the data source.

Edge computing originated in the field of media. It refers to the data processing mode that uses an open platform with sufficient computing power to provide services close to the data source. Of course, cloud computing can still access the history of

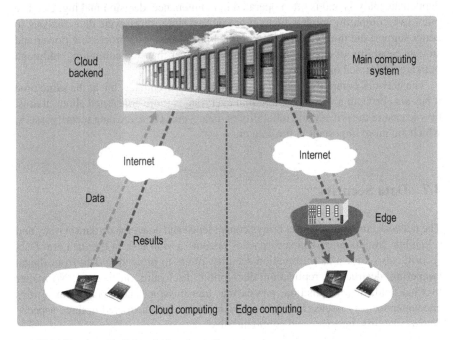

Fig. 4.13 Cloud computing and edge computing

edge computing data. With the development and application of new technologies, the demand for edge computing will continue increasing.

Cloud computing and edge computing are computing models whose fundamental algorithms are statistics and machine learning. They are based on the Internet and allow sharing of computing facilities, storage devices, and applications without geographical restrictions. Like clouds in the sky, we can see them when we look up, no matter where we are.

"Fog computing," a recently invented phrase, is a type of distributed computing that connects a cloud to some "peripheral" devices (for example, sensors). The term "fog" refers to the edge or perimeter of a cloud. Fog computing processes data in several centrally distributed nodes, known as "fog nodes," to make computing faster and more efficient. If cloud computing is about sending everything into the clouds, fog computing sends data into the fog around us. Its goal is to do as much processing as possible using computing units co-located with a data generating device to send processed rather than raw data, decreasing bandwidth needs.

There are three primary differences between cloud computing, edge computing, and fog computing, i.e., location of data processing, processing power, storage capabilities, and purposes. Cloud computing processes data on a central cloud server, provides superior and advanced processing technological capabilities, and stores more data than the other two computing models. It is best suited for long-term, in-depth data analysis and is often operated in maintenance, decision making, etc. The other two computing models focus more on real-time, short-cycle data analysis to better support the timely processing of local businesses. The processing power and storage capability of edge computing are even lower than fog computing, although both are performed on the devices/IoT sensor itself.

The network computing model has brought us convenience, but at the same time it has also brought about a problem that everyone is more concerned about, that is, how to ensure the privacy and reliability of data... etc. These are data security issues, which are more important in the data era.

4.7 Data Security

The network computing model brings convenience and concerns on data privacy and reliability. Maintaining data security has become a vital task in the data era. Data security, also known as information security, refers to protecting data from illegal tampering, destruction, replication, decryption, disclosure, misuse, etc. No matter in what field, data acquisition, processing, transmission, and storage can always encounter security issues (Fig. 4.14), which need to be considered in three aspects, data availability, integrity, and confidentiality.

Data availability refers to the property that authorized entities can access data and use on-demand. Data integrity ensures that data is not tampered with or can be quickly discovered after tampering during data transmission and storage. The confidentiality of data is the property of data not being obtained by illegal parties.

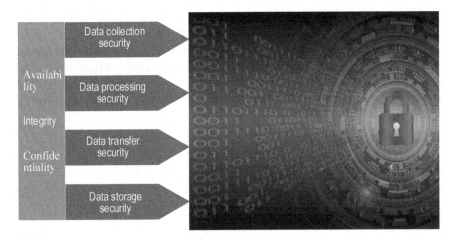

Fig. 4.14 Several aspects of data security

The above three aspects are the core issues to be considered when carrying out data security protection. By adopting various technical and management measures, data security ensures data availability, integrity, and confidentiality and allows the data system to run smoothly.

Data security has always been a classic and hot topic. Many factors can threaten data security, including hard disk damage, human error, hacker intrusion, virus, information theft, natural disaster, power failure, magnetic interference, etc. Anti-virus software used to be a technical measure to protect personal computers or individuals' data from damage. Data security concerns are no longer limited to the professional field; personal privacy and data confidentiality are common issues we all have to deal with daily.

With the development of network technology, all kinds of social software are attempting to master user information and record chatting, internet surfing, travel, shopping, and other information. In addition, the status of human activities and behaviors are predicted by big data. Without sophisticated protection technology and well-established protection laws and regulations, those personal behaviors and information might likely be leaked out. This is terrible as if someone is staring at you from behind!

Compared with the data acquisition and processing stage, the transmission and storage stage security is more serious. Data may also be attacked and gradually distorted in the transmission stage, and there is a risk of data leakage and tampering. The storage of various data types can lead to application conflict in the storage platform, causing data storage dislocation and management chaos.

The security of data must also not be ignored in engineering construction. Data leakage, tampering (data fraud), and other problems are often encountered during engineering data acquisition, processing, transmission, and storage. Intelligent construction should also conform to data security regulations. If there are problems with the training data, machine learning results will be unpredictable and have huge application risks.

The above is just a brief description of the importance of data security. However, preventing data fraud and achieving data security is under a different subject of information science, which requires more professional knowledge such as cryptography and legal issues. Interested readers can refer to books in those areas for more details.

Chapter 5
Understanding Perception Technology

Abstract This chapter starts with the methods of data acquisition, introduces the composition and main technical characteristics of perception technology, expounds on the critical issues of the application of perception technology in different fields, and analyzes the basis behind perception technology and the primary method of learning perception technology. A good understanding of engineering several sensing devices in the paper is briefly introduced. The difference and connection between automatic and intelligent sensing are explained. Finally, the engineering Internet of Things is briefly introduced, including the composition and critical technologies of the Internet of Things, cyber-physical system (CPS), the basic concept of the digital twin, etc.

5.1 How is the Data Obtained?

"Let data speak for itself" is a familiar phrase in many fields, including engineering. Data is the most critical resource in all areas. We have discussed data before but have not explained how the data (information) is obtained. Here we will briefly explain this topic.

Data has been with us since the start of human society, from simple counting in ancient times to online data recording. In the Internet era, massive data sources can be summarized into four types: personal behaviors, business behaviors, perceptual terminals, and data sharing, as shown in Fig. 5.1.

Data from personal behavior belongs to the category of personal socialization. Internet users have created a massive amount of social behavior data (such as social posts and data exchange records on mobile terminals), revealing people's behavioral characteristics and living habits, which has never happened before.

Data from business activities mainly include e-commerce and various payments, queries, evaluations, logistics, travel, etc. All kinds of e-commerce manufacturers often chase this part of data. Except for sharing, most network data are collected by web crawlers.

Data from perceptual terminals (the integration of sensors and data collectors) belong to the domain of expertise. Nowadays, many devices are equipped with

© China Railway Publishing House Co., Ltd. 2023
G. Xu and D. Wang, *Introduction to Intelligent Construction Technology of Transportation Infrastructure*, Springer Tracts in Civil Engineering,
https://doi.org/10.1007/978-3-031-13433-3_5

Personal behavior   Business activities

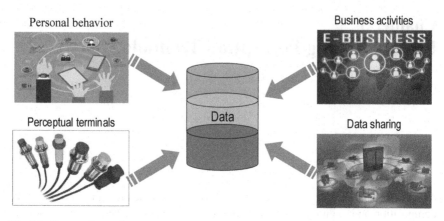

Perceptual terminals   Data   Data sharing

Fig. 5.1 Source of mass data

sensors to collect, monitor, and transmit information, which generates a variety of data and forms the foundation of the Internet of things.

Many data came from data sharing, such as music, photos, videos, and documents, which was impossible before the emergence of the Internet. In intelligent construction, sharing engineering data is crucial to break data isolation and encourage information circulation.

The above categories of data sources are not absolute but can overlap. For example, individual behavior such as running records may come from sensor data. In intelligent construction, data mainly come from different stages of engineering construction: survey, design, construction, quality control, management, and maintenance data. However, engineers have experienced a long journal in data acquisition, from manual operation to instrument measurement and automatic collection.

A review of the construction process of roads and railways shows that engineering data were firstly measured and recorded manually, such as topographic surveys, geological surveys, work logs, etc. Later, with the development of electronic information technology and computer technology, measurement and recording methods have changed, and electronic measurement and recording have become the mainstream. Some measurement and recording methods have realized automation and networking and are advancing toward intelligence, as shown in Fig. 5.2.

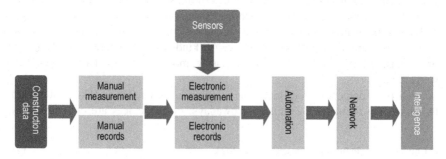

Fig. 5.2 Change in engineering data acquisition mode

Sensors have played an essential role during the advancement of data acquisition methodologies, especially in testing and detection. For intelligent construction, sensors are the primary technical means of data (information) acquisition, belonging to the scope of modern perception technology.

5.2 The Boom in Perception Technology

Human beings must rely on sensory organs to obtain information around them. Other means, including perception technology, are also needed as an extension to the direct sensory organs.

With the rise of the Internet of Things, perception technology is developing rapidly and has become one of the core technologies for acquiring information. Perception is an anthropomorphic term that reflects the desire to mimic a human being, meaning how the brain responds to information. The perception technology generally refers to all the technical means to obtain the target's (perceived object's) relevant information (i.e., attribute, identity, location), which can be regarded as the extension of sensing technology.

(1) Types of perception technologies

Perception technology is a general term for related technologies to obtain information, and its scope is still expanding. At present, the perception technology mainly includes Sensor and Transducer, Radio Frequency Identification (RFID), Global Navigation Satellite System (GNSS), Wireless Sensor Network (WSN), etc. In many cases, these technologies work together. Of course, AI should also be included as an indispensable technology for developing IntelliSense. Figure 5.3 summarizes the composition of all perception technologies in visual illustration.

The core of perception technology is the sensor, the primary technical means and hardware to obtain information. Perception, communication, and computer technology constitute the three pillars of information technology. Generally, the sensor does not work independently but is the most front-end component in the system. It

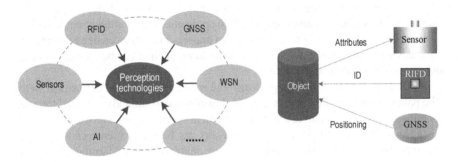

Fig. 5.3 Composition of perception technologies

senses the attribute of the tested object and converts it into electrical signals for the data collector to process, store and analyze. Sensors, such as force, displacement, acceleration, and temperature sensors, have a wide range of applications in engineering construction and are typically used together with data acquisition systems. New sensors, such as MEMS can be implanted with microprocessors and work independently to form tiny sensing terminals.

RFID, also known as electronic tagging, is a contactless technology that automatically identifies objects. Each RFID chip carries a unique code, issuing an "ID card" to an item. RFID consists of an electronic tag, a reader, and a computer. After affixed with an RFID tag, the item's relevant description should be established in the computer, corresponding to its code. When the reader does a contactless scan of the label, it can find information about the item. Such technologies include electronic toll collection (ETC), warehouse management in shopping malls, and automatic baggage sorting at airports. RFID is a future alternative to barcodes (one-dimensional and two-dimensional codes).

Global Navigation Satellite System (GNSS) is a standard technology found in mobile phones that provides accurate location information. The GNSS consists of three components—the space part has at least 24 satellites; the ground control system is responsible for monitoring the working state of the satellites; a user receiver uses the coordinates to determine a position. If using differential technology, its accuracy can reach the centimeter level. There are mainly four GNSS systems, GPS of the United States, GLONASS of Russia, Galileo of Europe, and COMPASS of China (Beidou).

WSN is a network composed of multiple sensors. Each node has a sensing terminal for sensing information and wireless data communication. They work together to perceive information about the environment or objects. WSN is now developing rapidly and receiving wide applications.

(2) The expanding scope of application

As the primary means of acquiring information, the application scope of perception technology is still expanding. Figure 5.4 shows only part of the application areas. In addition, perception technology is generally not applied alone but needs to be combined with other technologies such as the Internet of Things, data analysis technology, AI, etc. A few application cases are introduced briefly below to hopefully provide some inspiration and promote more applications in intelligent construction.

If perceptual terminals are embedded into household appliances like refrigerators and televisions in residential lives, network communication can be realized to carry out remote monitoring and control.

RFID can identify and manage drugs and medical devices in the medical field. The patient's identity and previous medical history can be quickly obtained through identification so that accurate and timely medical services can be provided.

In agricultural production, humidity sensors can be used to sense soil moisture. The information platform can carry out remote control of on-site farmland irrigation

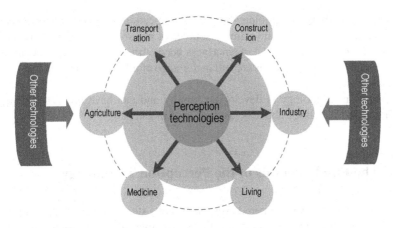

Fig. 5.4 Some applications of perception technology

according to the characteristics of crops. By applying location technology to agricultural machinery, the yield can be estimated. In addition, sensing chips implanted into cattle, sheep, and other animals can realize remote tracking and grazing.

In the industrial field, product manufacturing needs the support of a large amount of data and requires perception technology. In the intelligent factory, the real-time monitoring of production data, quality tracking, automated production, raw material management, workshop management, and many other activities need relevant data support and rely on various sensors and RFID to obtain data.

Satellite positioning technology provides critical location services for taxi management, bus management, hazardous goods transportation, and other applications. RFID is primarily used for the identification of various vehicles. By integrating sensing technology, positioning technology, RFID, network technology, and cloud computing, the interconnection between vehicles and vehicles and between vehicles and traffic facilities can be realized, forming a connected vehicle (CV) network. The prevalent unmanned driving is also an excellent example of integrating various sensing technologies.

Perception technology has always existed in engineering construction but is sometimes given a different name or terminology. Modern testing equipment and evaluation instruments often use sensing devices to realize automatic data collection, an embodiment of sensing technology. An "intelligent construction site" concept has recently emerged in pavement construction. During pavement material mixing, transporting, paving, and compacting, RFID helps identify material and construction equipment, and a satellite positioning system is implemented to monitor the roller compactor's operation coverage. Intelligent compaction is a technology that combines sensing, positioning, computer, AI, and other technologies. It can autonomously perceive the compaction quality information to conduct independent analysis and decision-making and take corresponding feedback control measures (see the next chapter).

In addition, sensing technology has many applications in urban construction and maintenance and is one of the primary sources of massive urban data. "Smart city" is a popular topic, but the level of smartness needs further consideration. Currently, most "Smart City" applications are still in the stage of informationization and require further exploration.

Perception technology is developing rapidly and can have broad applications. What we have discussed above is just the tip of the iceberg.

5.3   The Fundamentals of the Perception Technology

Perception technology has entered everyone's daily life. However, except for professionals, very few people are familiar with the principles of perception technology. In this section, let's look at some fundamentals of perception technology and discover how they work. This is especially crucial for those engaged in engineering and is developing and applying intelligent construction.

(1) RFID

As mentioned above, RFID consists of an electronic tag, a reader, and a computer. The working principle of RFID is shown in Fig. 5.5. When the labeled item enters the range of radio frequency signal (electromagnetic field) sent by the reader, it will receive the radio frequency signal, obtain the energy using the induced current, and send the coded information of the item to the reader for decoding. Then the reader will send it to the computer for processing. Timing refers to the working order between the reader and the tag.

RFID involves many technologies, including chips, electromagnetic fields, coupling components, control, radiofrequency, computer and decoding technology. The foundation behind these technologies is physics and mathematics.

(2) Positioning technology

Positioning refers to acquiring target position information using sound, light, and radio. Standard positioning technologies include satellite positioning, base station positioning, WIFI positioning, Bluetooth positioning, ultrasonic positioning, and UWB positioning.

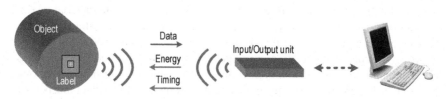

Fig. 5.5 The working principle of RFID

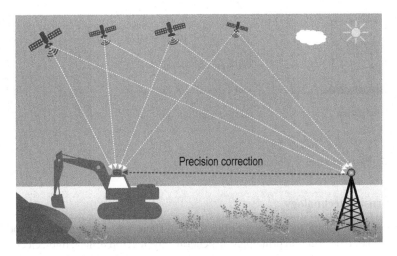

Fig. 5.6 The working principle of satellite positioning

Now let's take satellite positioning technology as an example to explain how it works. Generally, there are 24 satellites distributed evenly on six orbit surfaces, and more than four satellite signals can be observed anywhere in the world. The user's receiver calculates the coordinates of longitude, latitude, and height according to the satellite data. Differential corrections are used to improve the real-time positioning accuracy. After the correction, the positioning accuracy of the observation point (receiver) can be significantly improved. Figure 5.6 shows the working principle of the satellite positioning technology.

Real-time Kinematics (RTK) is a commonly used differential correction technology based on real-time carrier phase analysis between two observation stations. According to the precise coordinates of the reference station, the coordinate of the observation point can be corrected in real-time, and the accuracy can reach the centimeter level.

The principle of satellite positioning belongs to simple physics knowledge. The satellite continuously transmits the radio signal containing the position and the time. The distance from the satellite to the observation point can be calculated according to the time difference between emitting and receiving. Since the observation point (receiver) has three coordinates (latitude, longitude, and altitude), and a time error correction is needed, there are four unknowns. Therefore, four satellites are needed in the calculation. The most critical component in satellite positioning is using the atomic clock to achieve high precision in time calculation.

In addition to the physics theory, the hardware and software of the satellite positioning technology are also crucial, involving electronics, computer science, and mathematics (for analytical algorithm development).

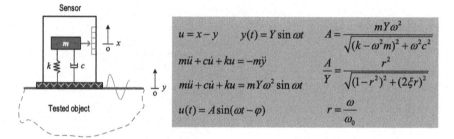

Fig. 5.7 Basic principle of vibration sensor

(3) Sensing technology

The core of sensing technology is the sensor. There are many different types of sensors based on different principles. In general, physics (force, heat, light, electricity, magnetic harmony, etc.), chemistry (chemical reactions), and biology (biological response) form the foundations of sensing technology.

The following briefly describes the working principle and the fundamental theory of the commonly used vibration sensors. Figure 5.7 shows that the vibration sensor is based on the vibration mechanics model.

The sensor records the relative motion u (displacement) between the mass block m and the measured object. According to the vibration equation, different types of sensors can be designed.

For example, assuming the frequency of the object being measured ω is much higher than the natural frequency of the sensor ω_0, i.e., $r \gg 1$. In that case, the sensor's displacement is consistent with the object's, $A \sim Y$, and the displacement sensor can be obtained.

Similarly, if the frequency of the object being measured ω is much smaller than the natural frequency ω_0, i.e., $r \to 0$, the sensor displacement is consistent with the acceleration of the measured object, and $A \sim \ddot{Y}$ can be deduced to obtain the acceleration sensor.

Different sensors can be obtained by changing the relative size of the mass, spring, and damping inside the sensor. This is the knowledge necessary for sensor design.

In many cases, the sensor's perceived information may not be enough or directly used to reflect the actual situation. Further data processing is needed, which often requires advanced techniques and will be discussed in other book series.

Through the above simple analysis, we can see that although the perception technology is very different, the common basis behind it is the basic discipline. Therefore, if you want to make original innovations, you must be proficient in physics and mathematics. This is not only true of perception technology, but also of many technologies. The so-called change is inseparable.

5.4   Ways to Learn Perception Techniques

The content of perception technology is vibrant and involves many subjects. How to learn and master this technology? It depends on the objectives of the learning. Take RFID as an example. General users can completely ignore the technical details and operate according to the user manuals, often referred to as a black box method. Developers need to understand the basic principles, be able to assemble chips and components into products, and write control programs. This is a grey box approach. The person who invented RFID must be familiar with all aspects of the technology, a white-box approach. These three learning methods have been introduced in Sect. 2.3 and applied to mastering any new technology, as illustrated in Fig. 5.8.

Using the grey box method, the technology developers engaged in intelligent construction should learn to master the sensing technology (including other new technologies). They should not only know the answer to the question but also understand the basic principles and know why it is the answer. This will allow them to develop or build hardware and software integration devices. For example, during the development of the new MEMS sensors, the author's team designed the circuit board for the purchased chip and wrote the control program to control the chip. This is the advantage of the grey box method. Adopting an utterly white-box approach to developing new technologies from scratch is unnecessary, although sometimes original innovation is still necessary.

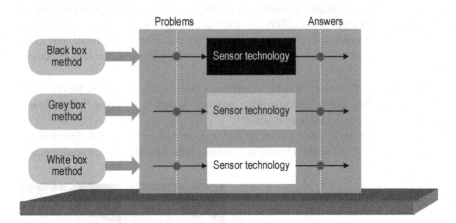

Fig. 5.8 Three ways to learn perceptual techniques

5.5   Sensoring Devices in Engineering

Sensors are the primary technical means to obtain all kinds of information. Sensing devices can be seen everywhere in engineering construction. For example, it is necessary to use various sensors during the construction stage to perceive various construction information such as material mix design, spreading thickness, compaction quality, etc. Therefore, the construction process can be fully monitored to achieve informatization, networking, and intelligentization. It is almost safe to say that without high-quality sensors, the informatization, networking, and intelligentization of construction will never be possible. Next, the authors will introduce some standard sensing devices required for intelligent construction.

(1)   The sensing devices in the design stage

The main task in the design stage is survey and design, and the sensing device focuses on geological and topographic surveys. Generally speaking, it can be divided into two categories: air perception and ground perception. Air perception is mainly based on remote sensing, while ground perception is the measurement using instruments equipped with sensors.

1.   Remote sensing sensors

Remote sensing is the acquiring of information from a distance. It uses sensors installed on remote sensing platforms (i.e., satellites, airplanes, etc.) to detect and record reflected or emitted electromagnetic energy produced by visible light, infrared, microwave, and other waves. Through photography and scanning, the electromagnetic wave information is processed to identify the nature and movement state of the objects on the surface. Figure 5.9 explains the working process and remote sensing results.

The sensors in remote sensing technology are mainly based on photoelectric effects, including optical photography sensors, scanning imaging sensors, radar imaging sensors, and non-image sensors.

The multispectral sensor can scan the surface, collect data in different wavelength ranges, and generate a two-dimensional surface image. An infrared sensor is

Fig. 5.9   The working process and achievements of remote sensing

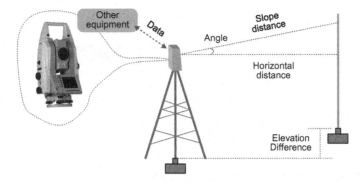

Fig. 5.10 Function of total station

responsible for collecting the thermal radiation information of the object (invisible to human eyes) and generating the image of the radiation temperature distribution of the object. Microwave sensors can detect microwave energy under various environmental conditions and collect data.

These sensors are mainly related to electromagnetic wave theory, which is relatively easy to understand. Still, the development of the sensor products is not easy, given their complex structures. Special agencies generally carry out image acquisition. People engaged in engineering construction should be familiar with the characteristics of these sensors, select appropriate sensors according to needs, and learn to apply the remote sensing images. In addition, methodologies such as AI technology should be considered for remote sensing data processing.

2. **Instrument sensors**

The topographic survey uses three main instruments: level, theodolite, and Total Station. Figure 5.10 shows a brief introduction to the sensors in the Total Station, which are mainly based on the photoelectric effect.

Total Station is the abbreviation of Total Station Electronic Rangefinder. The measuring equipment integrates the measuring functions of Angle (horizontal Angle and vertical Angle), distance (slope distance and horizontal distance), height difference, and microcomputer processing. The measuring results can be displayed automatically, exchanging data with peripheral equipment. All the measuring jobs at the target station can be completed by placing the instrument simultaneously. At present, the new total station can automatically complete the identification and measurement of multiple targets under the condition of unmanned intervention, which is called a "measurement robot." Interested readers can refer to the relevant information.

There are many sensors in the total station, each undertaking different tasks. For example, a capacitive sensor is for leveling (bubble) calibration; a fiber optic sensor is for measuring distance; a grating sensor is used for angle measurement, and CCD (Charge Coupled Device) is the "eye" of the total station, which does the scanning.

The total station is used in surveying work, which can simultaneously carry out control and topographic surveys. It has also been successfully applied to install and

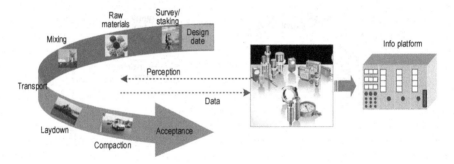

Fig. 5.11 Road construction process and information perception

debug industrial equipment and components, observe building deformation, conduct hull construction, monitor geological hazards, etc. The total station is believed to play an essential role in intelligent construction, and the sensor is its critical sensing device.

(2) **The perception device in the construction stage**

The construction stage is the main battlefield of intelligent construction, which relies on various perception devices to obtain a large amount of data. As illustrated in Fig. 5.11, road construction processes, such as construction lofting, material mixing, paving, rolling, and testing, are mechanized or automated with various measurement and control equipment. Corresponding sensors are built into the equipment to perceive construction information (data).

Perception devices are included in different instruments and equipment for data acquisition during construction. Next, I would like to introduce some popular perception devices.

Optical induction sensors, often used in measuring instruments, are based on the principle of light reflection and use the speed of light and travel time to determine the distance. The leveling sensor used in pavers belongs to this type. The non-contact thermal infrared temperature sensor is widely used in asphalt pavement construction. It is based on the infrared radiation principle of the band located at 0.75–100 μm. It can measure objects' motion speed, surface temperature, and temperature distribution and be used for thermal infrared imaging.

The most commonly used sensing devices in engineering construction are mechanical sensors, including force, displacement, and acceleration. They are used in various mechanical tests, field monitoring, and intelligent compaction. Each sensing unit can follow different mechanisms. For example, force sensors can be divided into piezoresistive, piezoelectric, inductive, etc. Using some sensing units, force information can be converted to electrical signals. Mechanical sensors are widely used in engineering, and readers can refer to relevant information for a more detailed explanation.

Regardless of sensor types, the measurement is converted to a digital quantity to be recognized by the computers. The basic process is shown in Fig. 5.12.

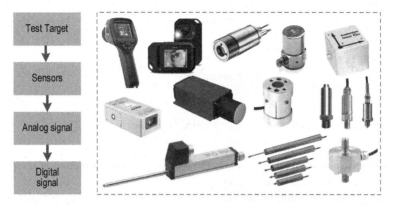

Fig. 5.12 The process of being measured into digital quantity and the diagram of the sensor

In the construction stage, in addition to these sensing devices, ultrasonic sensors and various measurement sensors are often used for mixing quality control at the plant, and technologies such as satellite positioning and RFID have also been applied.

(3) **The sensing device in the maintenance stage**

The maintenance stage's main task is collecting all kinds of data to serve the maintenance decision. There are sensing devices on all kinds of detection instruments, such as force sensor and acceleration sensor on the falling weight deflectometer (FWD), laser sensor, ultrasonic sensor, and electromagnetic wave sensing device on the comprehensive detection vehicle of Ground Penetration Radar (GPR), and sensors embedded in the structures.

At present, almost all sensing devices are developed in the direction of miniaturization, digitalization, and intelligence. Improving the level of intelligence will especially be more favorable for non-professionals.

Sensors are the fundamental but critical products for the development of modern technology. They are recognized as a promising industry attracting worldwide attention for their high technical content, good economic benefits, and broad market prospects. We should pay close attention to the development trend of new sensors to introduce them into intelligent construction.

5.6 Automatic Sensing and IntelliSense

With the rapid development of sensing technology, new sensors have been developed with built-in single-chip microcomputers such as the Micro-Electro-Mechanical Systems (MEMS) sensors. Concepts such as automatic sensing and intelligent sensing have also appeared. Still, they have been misunderstood in many cases, and it is necessary to clarify the differences and connections.

The difference between automation and intelligence can be described using a simple principle, which has also been discussed in Sect. 1.5. For a job, if the machine

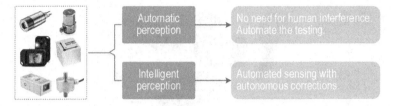

Fig. 5.13 Significant differences between automatic and intelligent perception

replaces human manual labor, it belongs to the category of automation; if it replaces mental work, it belongs to intelligence.

The main feature of automatic sensing for sensors is automatically completing the measurement work without human intervention, while intelligent sensing is based on automatic sensing. The sensor can perform autonomous learning, analysis, reasoning, judgment, and decisions based on the collected data and independently correct errors, as shown in Fig. 5.13.

Currently, most of the intelligence we are exposed to is just advanced automation, not too much intelligence, and is far from the real wisdom! However, there is no need to obsess over these concepts as long as they serve intelligent construction.

5.7 Internet of Things in Engineering

So far, we have had some basic knowledge about perception technology. One of the primary uses of perception technology is to build the Internet of Things in engineering. Next, let's briefly discuss this issue in conjunction with intelligent construction.

(1) **What is the Internet of Things?**

Suppose the sensing device is implanted into various items so each item can sense and connect to the network. In that case, the items will form a network that can exchange information. This is the popular definition of the Internet of Things (IoT). IoT allows everything to be connected by the Internet (Fig. 5.14), the new stage of internet development.

The early-stage IoT is based on the logistics system and uses RFID as a substitute for the barcode to realize the information management of goods. Now the IoT is an essential part of modern information technology. Its connotation is undergoing profound changes, not only to be connected to things but also to carry out real-time control.

(2) **Composition and key technologies of IoT**

From the technical point of view, the IoT can be divided into three layers—the perception layer, the network layer, and the application layer, as shown in Fig. 5.15. They constitute the essential elements of the Internet of Things.

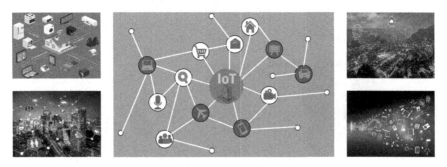

Fig. 5.14 Things are connected to form the Internet of Things

Fig. 5.15 Composition of the internet of things

The perception layer is mainly responsible for "perceiving" events and data in the physical world. It is the key to the information collection, involving sensors, RFID, positioning technology, and network information collection.

The network layer is also called the transmission layer or the communication layer. It is responsible for the safe and reliable transmission of the sensed information. It is a high degree of integration of sensor networks, wireless communication technologies, and network technologies.

The application layer (including control) faces various fields such as industry, transportation, construction, agriculture, medical care, etc. The information platform is responsible for calculating and processing the perceived data and coordinating, sharing, and intercommunicating information. It will ultimately realize real-time control, precise management, and scientific decision-making of items and provide professional-level services for various fields.

The Internet of Things is based on many technologies. RFID is used to identify the identity of items. Sensors are used to sense the attributes of items. Positioning technology is used to determine the location of items. Wireless communication includes voice and data networks and near-field communication technology such as Bluetooth,

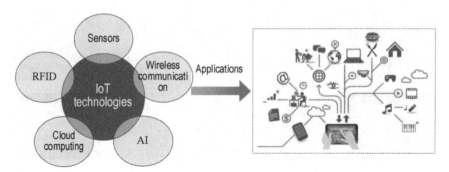

Fig. 5.16 Key technologies of the internet of things

ZigBee, etc. It is the communication medium between items and people and between items and items. AI is responsible for information processing, analysis, and decision-making. Cloud computing realizes the storage and calculation of massive data. These technologies will be integrated into the information platform of the Internet of Things and thus applied to various fields, as shown in Fig. 5.16.

(3) **What can the Internet of Things do?**

The Internet of Things, also known as the "sensor network," provides the interconnection of items, has data processing capabilities, and can implement items' necessary management and control.

For example, by embedding sensing terminals into various items in the fields of roads, railways, airports, buildings, dams, municipalities, etc., and combining them with the Internet, human society and physical systems can be integrated. The corresponding information model can thus be established in computers to realize real-time management and control of personnel, materials, equipment, quality, and progress.

The application of the Internet of Things can be divided into monitoring type (such as logistics monitoring and pollution monitoring), query type (such as intelligent retrieval and remote meter reading), control type (such as intelligent transportation, smart home, and street light control), and scanning type (such as mobile wallet, No parking fee), etc. For intelligent construction, monitoring and control will take the lead.

The Internet of Things can manage and control the items in the network and enable us to manage production and life in a more refined and dynamic way. It will optimize resource utilization and productivity and improve the relationship between man and nature. Therefore, the Internet of Things is considered the third wave of the information industry after computers and the Internet.

(4) **IoT in Engineering**

Building the Internet of Things in engineering for intelligent construction is necessary. The intelligent construction information platform mentioned in Sect. 2.7 is the embryonic form of the Internet of Things in engineering. Still, it is not yet fully equipped with the ability to manage and control each item in real-time, and it still needs to be developed and improved.

Internet of Things in Engineering is one of the critical technologies of intelligent construction. It is also the primary platform to realize the management and control of the whole life cycle of engineering construction. To build such a system, at least the following requirements must be met:

1. All objects should have perception capability and network connectivity

Having perception and network connection capabilities for all objects is the most basic requirement of the Internet of Things. It requires that various items be equipped with corresponding perception terminals.

For example, in the construction phase, a paver could use RFID to identify its identity; a positioning system could be equipped to determine the paving position; a speed sensor or satellite positioning technology could be applied to control the paving speed, and a laser sensor could be used to determine the paving thickness and leveling control. RFID needs to be configured to identify identity; an intelligent compaction control system is configured to control compaction quality. In addition, the paver, the roller compactor, the plant, management, and inspection, should be interconnected, and all information should be transmitted to the platform.

2. Have smooth and safe wireless communication capability

In engineering construction, items such as materials and equipment are in dynamic change. Therefore, it is necessary to have a smooth and secure wireless communication capability. With the popularity of 4G and the beginning of 5G (and 6G, etc.), wireless communication is no longer a bottleneck that hinders the development of the Internet of Things.

3. Capable of real-time management and control

Real-time management and control are advanced requirements for the Internet of Things in engineering. It is required that the information platform and engineering construction should be able to interact to realize real-time management and control of the site.

In addition, to ensure the smooth implementation of the Internet of Things in engineering, many network-related technologies, such as firewalls, encryption, and secure access, also have technical requirements and must be maintained by professionals.

Internet of Things in Engineering is a complex dynamic management and control system covering design, construction, and maintenance stages. Various resources can be integrated to facilitate unified management and control through the information platform. Figure 5.17 shows the basic framework of the Internet of Things during the construction phase.

The construction phase is the home field of intelligent construction. It is also the place where many technologies are applied and displayed. Through the Internet of Things, the information of each aspect of the construction process can be linked, and comprehensive management and control can be realized on the information platform.

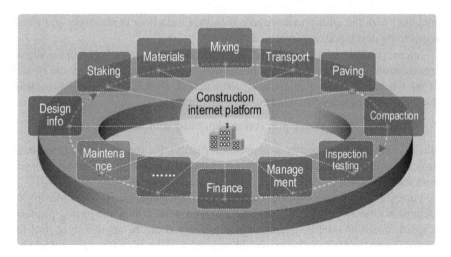

Fig. 5.17 The basic framework of IoT in the construction stage

Internet of Things in engineering is an upgrade of the current information platform. Not only does it have the ability to interconnect and communicate with each other, but more importantly, it can perform real-time management and control based on the perceived information. Although this desire has not been fully realized, most have been automated and have a good foundation.

(5) **Cyber-physical system—CPS**

The Internet of Things is linked to another concept: Cyber-Physical systems (CPS). CPS is a complex network system connecting the physical and information worlds. It connects physical devices to the Internet and reproduces the physical world in the computer through information (data) interaction. It has the capabilities of calculation, communication, control, coordination, and self-management. Finally, it achieves a high degree of integration between the real physical world and the virtual digital world (Fig. 5.18).

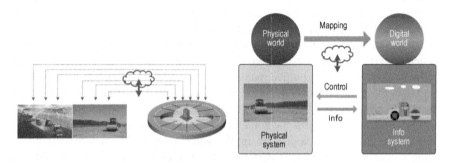

Fig. 5.18 Information physical system (CPS)

CPS can be regarded as a new stage of the Internet of Things development or new development of computer simulation technology. Still, it is different from the conventional Internet of Things or simulation technology because it has feedback control capabilities. A similar concept is "Digital Twin," which may be easier to understand.

Physical equipment generally refers to all objects in the natural world, the "items" in the Internet of Things. The physical world is called the real world, including various physical systems and human society. The information world is called the virtual or digital world, mapping the natural world in the computer or virtual representation. They are connected through data (information).

With the development of modern information technology, scientific research methods are also changing. A method of moving the real world into a computer for simulation has been rapidly developed, called computer simulation technology or digital simulation technology (virtual reality technology also belongs to this category). Digital simulation technology is a commonly used method in engineering construction, and the famous Building Information Modeling (BIM) is one of the representatives of this type of method. CPS can perform virtual simulations of physical equipment and implement the remote operation. These functions are not available in standard computer simulation technology and are unsuitable for the current Internet of Things.

In the eyes of computers, there are two worlds: a physical world (such as engineering construction and mechanical manufacturing) composed of various entities and people. The other is a virtual world composed of digital models. CPS is the bridge and link between these two worlds. It uses computers to integrate perception, communication, and control technologies and uses the network to operate physical entities to realize real-time perception, analysis, decision-making, and dynamics management. CPS can connect the entire world and will change our interaction with the physical world.

The current network layer, whether the Internet of Things or CPS, has matured. The development of the perception layer is also very rapid. However, the application control layer is still relatively backward. The application layer directly provides users with specific services, and the information platform is its primary form of expression. It has the closest connection with engineering construction and has excellent potential for future development. For engineering construction, the construction of a core platform with information management has been completed. A platform with interconnection as the core (Internet of Things) is emerging. The research and development of a platform (CPS) with remote control should focus on intelligent construction.

Chapter 6
Intelligent Engineering Construction

Abstract This chapter starts with the essential characteristics of intelligent construction "perception, analysis, decision-making, execution." It firstly analyzes the feasibility of intelligent technology application in the design stage and the critical problems to be solved. It then focuses on the intelligent problems in the construction stage. Including the intelligentization of the service guarantee system and the intelligentization of each link in the construction process, the main tasks of intelligent maintenance are analyzed, the feasibility of intelligent construction management technology is analyzed, and the basic concept of virtual construction is introduced in this basis. It also provides some main contents of virtual simulation of road engineering. Finally, it introduces the application case of intelligent construction—the implementation process of intelligent compaction and related machine learning algorithms. In addition, the risks of intelligent construction are also discussed.

6.1 The Dawn of Intelligent Construction

With the rapid development of information technology, transportation infrastructure has entered a new era of intelligent construction. Information technology and the internet have been integrated with construction at survey and design, construction and production, or maintenance stages.

(1) From informatization to intelligence

Intelligent construction and information technology are closely related to each other. Practice in other industries has proved that intelligence heavily relies on informatization and networking, and the engineering industry has no exception. Informatization and networking are the basis for developing intelligent construction, as shown in Fig. 6.1.

The computer is one of the main tools for implementing intelligent construction. In the early 1990s, information technology began to be applied in the construction industry. One example of tremendous success is the replacement of traditional manual drawing with computer-aided drawing (CAD). Since the beginning of the

© China Railway Publishing House Co., Ltd. 2023
G. Xu and D. Wang, *Introduction to Intelligent Construction Technology of Transportation Infrastructure*, Springer Tracts in Civil Engineering, https://doi.org/10.1007/978-3-031-13433-3_6

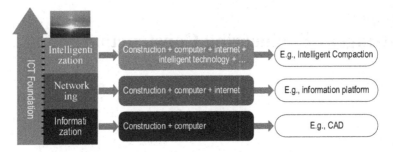

Fig. 6.1 Fundamentals of intelligent construction

twenty-first century, the acquisition, use, control, and sharing of information have become more accessible and cheaper due to the extensive use of the Internet. In the Internet and data era, sharing information has become popular. The development of an information platform is an example. The popularity of the Internet has also promoted the development of the Internet-of-Things and CPS and further promoted the combination of information technology and the construction industry, which are the essential foundations for the development of intelligent construction.

The informatization (another name is digital) and networking of engineering construction have also promoted sensing technology and machine learning in engineering. Intelligent construction technologies have been implemented, among which intelligent compaction technology has developed the fastest. It has the ability of autonomous "perception, analysis, decision-making, and execution," which has brought a brighter future to intelligent construction.

(2) **Implementation steps of intelligent construction and main tasks at each stage**

The four essential characteristics of intelligent construction, namely "*P*erception, *A*nalysis, *D*ecision-making, and *E*xecution," are the basic steps for the implementation. According to practical experience, it is not difficult to implement these steps. Figure 6.2 gives a detailed description of the steps of intelligent construction. The key is training the machine to learn and determining what knowledge to learn.

The detailed contents for the implementation of intelligent construction vary at different stages. Table 6.1 shows the major implementation components in the design, construction, and maintenance stages, which need to be modified in operation.

According to the survey, some technologies involved in intelligent construction have been widely used in other industries. Integrating intelligent construction with actual construction practice in depth is a question to solve in further work.

"Perception" in intelligent construction relies on sensing terminals and automatic measurement with computers as the core. Sensing devices generally do not need to be developed separately, and more research is focused on data processing, which needs to be modified according to field knowledge. For example, at the survey and design stage, remote sensing equipment focuses on interpreting the information contained in the image. The interaction between construction machinery (paver, road roller,

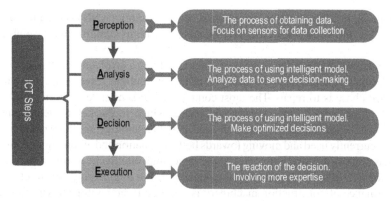

Fig. 6.2 Basic steps of intelligent construction implementation

Table 6.1 Major component of intelligent construction at each stage

Characteristics	Stages		
	Design	Construction	Maintenance
Perception	Test, survey data	Material, construction, and QA data	QA data
Learning	Design capability	Construction technology	Decision capability
Analysis	Related data	Related data	Related data
Decision	Optimized design	Optimize construction technology, QC	Maintenance priority
Execution	Blueprint	Construction	Maintenance

etc.) and materials is easy to be perceived in the construction stage. The key is to relate the collected information to construction quality. The experience has proved that these problems are not difficult to be solved with a solid professional foundation.

"Learning, analyzing, and decision-making" is the problem that AI needs to solve. In the past, most of the applications of machine learning algorithms in engineering include prediction by replacing traditional regression models. Intelligent construction aims to use the machine to learn and use professional knowledge, such as design skills, construction technology, and maintenance strategy, to assist human work. This professional knowledge could be deterministic, uncertain, or possibly vague in a non-mathematical form. Based on the current development of AI, it should be feasible to allow machines to study and use professional knowledge (building a knowledge-based intelligent system is called knowledge engineering).

For example, you can first transform professional knowledge and experience into several "rules" to form a knowledge base and then build a new type of expert system to conduct this task. A new type of expert system is a hybrid intelligent system that combines machine learning algorithms, such as neuro-expert system, neuro-fuzzy system, evolution-neural network, fuzzy-evolution system, etc. They can exert their strengths, enhance learning ability, and deal with the knowledge with uncertainty and ambiguity. For example, an artificial neural network (ANN) is used to train the "rules" in the knowledge base, and genetic algorithms optimize the "weights" of ANN.

Regarding "execution," more professional fields are involved. The "execution" in the design phase produces construction drawings, and the "execution" in the maintenance phase is to repair. The most complex and essential "execution" task is the construction work in the construction phase, which is the key to generating the "product." With the development of science and technology, mechanized construction is currently used and moving towards both automation and autonomous construction, becoming an attractive topic. At the same time, intelligent construction is in the development phase, like intelligent compaction. For engineering construction, the performance of construction machinery is required to be intelligent with functions to facilitate intelligent construction.

Currently, intelligent construction is implemented in the construction phase. Intelligent construction machinery will replace massive labor and part of the human brain, enabling humans to work more creatively. At the same time, intelligent construction will effectively reduce the consumption and waste of resources and energy, leading to the green development of the construction industry.

6.2   The Feasibility of Intelligent Engineering Design

There are two main tasks in the design stage: to obtain all information needed for the design and to design according to technical standards. They are also the basis of intelligent engineering design.

(1) **The foundation of Intelligent engineering design has been initially established**

Intelligent engineering design allows the computer to carry out independent design like a human engineer in the whole process or critical design processes based on the relevant information it perceives, achieve goal optimization, and generate construction drawings that meet the requirements. The following describes the current state of practice of the essential characteristics of intelligent construction, namely "perception, analysis, decision-making, and execution."

1. **Information perception is not a problem**

With the development of perception technology, obtaining information such as topography, landform, and geology is no longer a problem. Most information can be collected through automatic sensing, and the amount of data has far exceeded the previous magnitude. The focus is switched to the application, and the further improvement of perception technology will rely on developing other disciplines.

2. Automatic drawing and visualization are easy to implement

Automatic drawing and visualization are part of the "execution," allowing the computer to automatically draw and display the results (including three-dimensional visualization and dynamics) according to requirements. The computer will automatically generate charts and graphs according to the engineer's requirements. Therefore, automation and visualization are very easy for CAD, but it is tricky to train the computers to learn and use the design skills in intelligentization.

Intelligent engineering design is not necessarily automatic drawing. The key is to see whether the computer can propose a design plan and make an optimal decision to replace part of the engineers' work.

3. Machines can learn and master design skills

Training computers to learn and apply design skills is one of the cores of intelligent engineering design. Learning, analysis, and decision-making are closely related to humans' need to learn (train) and work.

We know that the design skills of engineers are mastered through learning existing design experiences, design theories, and technical standards (of course, a lot of practice is required). The first thing for intelligent engineering design is to train the machine to learn and master this knowledge. A new type of expert system is one solution, as shown in Fig. 6.3.

The expert system can process knowledge and be implemented in many industries, but it lacks learning ability. Other technologies need to be introduced to develop a new intelligent engineering design system to solve this problem. For example, ANN can train models for the design system.

Our task is to split the design-related knowledge into several rules and store them in the database. This requires meticulous work and close cooperation between professional engineers, computer programmers, and users. Mature experience in other disciplines can be learned. For example, classification or clustering algorithms in machine learning can be applied for data pre-processing.

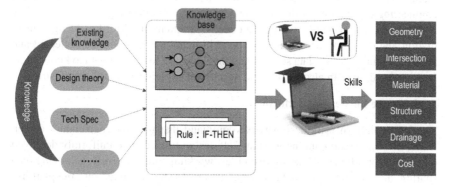

Fig. 6.3 Flowchart of training computers to learn and use design skills

Combining ANN (or other AI technologies) with expert systems is a new development trend. In an expert system, knowledge is manually converted into rules. The relationship, suggestion, instruction, or strategy expressed by rules exists independently. In ANN, rules are generated through learning from data, and knowledge is stored in the weights of neurons in the network. An ANN can include the related knowledge, automatically obtain new weight distributions, and update the knowledge based on new training data.

The use of ANN mainly involves classification problems. At present, ANN has dozens of models with different learning algorithms. It is necessary to select a suitable model according to the specific situation.

ANN requires that the knowledge is converted into digital data. Still, some knowledge may be challenging to digitize, making ANN not feasible. In this case, other machine learning algorithms need to be considered accordingly. In addition, the uncertainty of knowledge should be considered using the probability method, and the unclear knowledge can be processed using the fuzzy method. In short, AI technology needs to be applied flexibly, without being restricted to a specific form, which is also the idea of applying innovation.

(2) **Ideas for developing intelligent engineering design**

Design work can be conducted once the machine has mastered the design skills. The essence of design is continuously optimizing the plan and maximizing the satisfaction of the established goals, known as the "decision-making" process. The foundation of intelligent engineering design under the new situation (information age, data age, and intelligent age) has been preliminarily formed, but how to carry out intelligent engineering design work still needs to be discussed and considered. Here is a brief talk about ideas.

1. **Build a knowledge base**

Engineering design involves complex system engineering, and design issues should be viewed from a systematic perspective. An essential task for developing intelligent engineering design is building the knowledge base required for intelligent engineering design, one of the "big data" sources of engineering.

The summarization of existing data needs our attention. Prior knowledge, information, data, and experience from experts in many fields should be collected and utilized. Figure 6.4 lists data sources and the flowchart to convert data to knowledge. In this regard, we can also learn about successful experiences in other industries.

2. **Improve the design platform**

The design platform is the design work's command, dispatch, and coordination center. Using informatization and networking, all design tasks can be uniformly deployed on the platform by coordinating resources, working collaboratively, achieving data sharing, and breaking the "data islands." It can also be integrated into the platform (three-dimensional, virtual, and dynamic) to show the design.

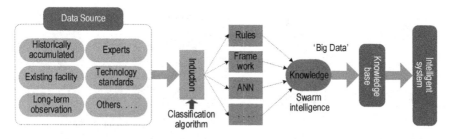

Fig. 6.4 Flowchart from data to knowledge

Many design departments are equipped with office automation (OA) systems at the design stage. What needs to be done is to supplement and improve them to form an independent design platform (or with additional development) and gradually move to an intelligent engineering design platform, becoming a part of the intelligent construction information platform.

3. **Breakthrough from point and surface**

The design can be complicated and straightforward if it is a road or railway project. There are many factors to consider, such as the structure must carry the live loads at different speeds and adapt to the natural environment. However, the structure materials are challenging to be controlled entirely; and the interaction between the structures is unclear. How to consider these issues in the design has always been difficult.

Intelligent engineering design will be conducted for the entire life cycle. With the support of various emerging technologies, investigating the best design plan through in-depth mining of massive data is not guaranteed to solve all problems.

In the 1990s, the development of engineering design expert systems became a popular direction with the rise of CAD. Still, after several decades most results have not yet been found (at least in the road and railway fields). What is the problem? The answers will be valuable references for the development of Intelligent engineering design.

Intelligent engineering design is still a new thing that is difficult to implement—developing innovative strategies, solving the key issues, and identifying the appropriate intelligent technology to make breakthroughs.

(3) **Key points of intelligent design for roadway**

Since the alignment of roads is more complicated than railways, the following is based on an example of road engineering. The main content of road engineering design is shown in Fig. 6.5.

It is not difficult to make a design plan, but it is not easy to make an excellent one. After CAD, it is not difficult to apply automatic drawing with a computer. The difficulty is how to obtain an excellent solution (optimization problem).

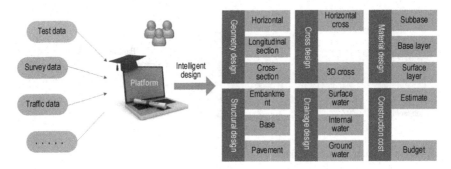

Fig. 6.5 Flowchart of roadway design

1. Main points of intelligent design for alignment

The current road alignment design considers the horizontal and vertical sections separately, later synthesized into a spatial form. This will neglect some interactive features. The design process needs to be done repeatedly until the kinematic and dynamic requirements of the vehicle are met. New design methods need to be studied, and three-dimensional design should be one of the development directions.

Given the current practice, intelligent engineering design should first solve the evaluation problem of the design plan, propose suggestions and modifications (optimization of design alternatives), and gradually expand to the entire design process.

As shown in Fig. 6.6, there are multiple choices of route alignments from the start to the end of the route. Design is an optimization process of constant modification through trial and error. This process can be handed over to the computer, allowing the machine to evaluate the preliminary design scheme, pointing out the advantages and disadvantages, and then making improvements until the requirements are satisfied.

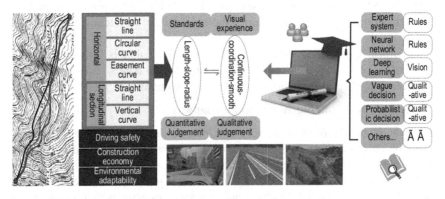

Fig. 6.6 Intelligent evaluation and optimization

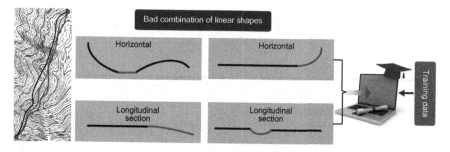

Fig. 6.7 Intelligent recognition of undesirable combinations of linear and curved segments in roadway alignment

Figure 6.6 shows the main contents of roadway alignment using intelligent evaluation and optimization, including the alignment's length, slope, and curve radius and the continuity, coordination, and smoothness of curves.

Horizontal and longitudinal alignment designs determine the curve's length, slope, and radius. There are precise requirements in the technical standards (for example, when the car speed is 120 km/h, the maximum longitudinal slope is not more than 3%, and the maximum length is not more than 900 m), which are quantitative judgments. It can be evaluated and optimized using a neuro-expert system. The requirements in the technical standards can be transformed into several rules, some of which can be expressed by ANN.

The wrong combination of linear and curved segments in roadway alignment (Fig. 6.7) will directly affect the driver's visual experience and driving safety. It needs to be comprehensively considered from the aspects of driving safety, cost y, and environmental adaptability. The curves must be continuous, coordinated, and smooth from the driver's perspective. This part is mainly based on the qualitative judgment from the visual angle of the vehicle driver. Image recognition technology can be utilized for evaluation.

For example, deep learning technologies can be applied to judge, classify and make decisions based on images or videos. The premise is to prepare various images or videos (training data), train the machine to learn, and master the evaluation technology., Fuzzy and probabilistic methods can be used for qualitative description.

The above is an assessment from the perspective of driving safety. As for the economics, the costs of different design alternatives need to be compared. For environmental adaptability, engineers shall select the route based on the principle of "environmentally friendly and resource-saving" and pay attention to the balance of embankment filling and excavation, which can also be used in intelligent technology.

2. Key points of intelligent design for the structure

The linear design is mainly based on the kinematics of the vehicle. For roads or railways, the structural design is based on the vehicle's dynamic characteristics. Although the structure is simple (Fig. 6.8), mechanistic analysis is complicated. The

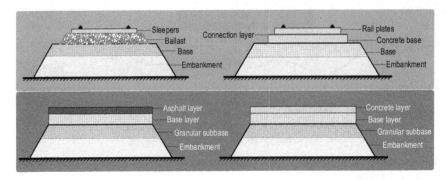

Fig. 6.8 Basic components of railway and road structures

current structural design is mainly based on experiences, and mechanistic analysis is only considered supplemental.

> There are many reasons for the problems in the mechanical calculation, in addition to the inconsistency with the fundamental assumptions. First, the design is solved by statics, but the actual load is dynamic; second, the constitutive relationship does not conform to Hooke's law; third, the physical parameters are estimated; fourth, water and temperature are rarely considered. Many current calculation results are inaccurate and can only be used as a reference for these reasons. It is worth studying whether it can be determined through prior knowledge (data) and intelligent technology.

The basic structure of the railway is "subgrade + track," which is divided into two types: ballasted track and ballastless track. The upper part of the subgrade of the ballastless track is a graded gravel layer, which is mainly used for high-speed railways. The basic structure of the road is "subgrade + base + surface." The surface layer can be divided into flexible (asphalt concrete pavement) and rigid (cement concrete pavement). The base layer can be semi-rigid (lime-fly ash stabilized aggregate) or flexible (graded gravel). Let's take road engineering as an example to discuss the intelligent design of pavement structures.

Structural design (also called pavement design) needs to determine the structure type of the surface layer and base layer, calculate the thickness of each structural layer, and select the material of each structural layer. In addition, the subgrade, base, and surface layer as a whole (system) need to be optimized to reach the maximized economic benefit.

Currently, the road structure is relatively simple without many options, and the design is mainly based on experience. The thickness of the structural layer is not directly calculated. Instead, the layer thickness is first assumed and then checked (primarily based on experiences), and the required modulus is also determined based on experiences. Therefore, it is essential to accumulate large amounts of historical data.

Based on the above analysis, with prior knowledge and experience as the core, building a knowledge base for the structural design of roadways should be the start of intelligence. The knowledge base includes the existing design data, traffic load, environment, long-term observations, maintenance data, etc. (see Figs. 6.3 and 6.4). The knowledge can be expressed in IF–THEN or considered to use ANN to express some knowledge (rules).

It is more realistic to develop and optimize the structural design of roadways based on massive data (knowledge). That is, Intelligent engineering design driven by big data, which is also applicable to the following material design. The design method based on mechanistic analysis still needs in-depth study.

3. Key points of Intelligent design for material

The materials mentioned here mainly refer to the various raw materials that constitute the base and surface layers of the roadway. The materials are crucial elements of the road structure (Sect. 2.4). After paving and rolling, the pavement structure is finally formed. Material design mainly determines the composition and content of the mixture, including the gradation of aggregates and the amount of binder (such as cement, asphalt, etc.).

The material design should also include the manufacturing process (construction/production), which affects pavement performance. For example, even if the aggregate gradation meets the requirements, it is not easy to roll it into a dense structure if the compaction parameters are unsuitable. This is why adding cement stabilizer in the aggregate layer for high-speed railway. Therefore, the construction process should be considered in material design. Like cooking, if there are only raw materials and recipes but no production methods (craftsmanship), it is impossible to make delicious food.

The material design is a mixture design. The purpose is to produce a mixture that meets the structure requirements. However, it is difficult to directly derive the relationship between the mix components and the structural performance, which can only be obtained through experiments. It can be seen that the material design is based on accumulated experience and the law obtained from the experiment. Therefore, organizing the existing data into a knowledge base and using an expert system is natural.

Figure 6.9 shows the method of building a neuro-expert system for material design. ANN generates "materials ~ process ~ performance" rules, where IF–THEN rules and frameworks can describe various raw materials' types, compositions, particle sizes, shapes, gradations, and implementation processes.

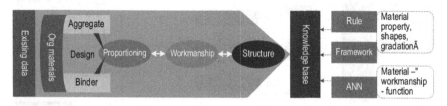

Fig. 6.9 Neuro-expert system designed with building materials

The ideal neuro-expert system should have the following functions: according to the performance requirements of the structure and the rules in the knowledge base and the information in the database, the expert system can infer the material type, mixing ratio, and process; according to the given material types, mixing ratios, and process parameters, the expert system can use the information in the database and the rules in the knowledge base to deduce the performance requirements (performance prediction) that the designed structure can satisfy.

4. **Integrating structure, material, and construction**

The structural design allows the structure to undertake the effects of traffic loads, temperature, and humidity. The material design allows the raw materials to be mixed, paved, and rolled into the pavement structure. The construction process is to process the mixture into an engineered structure. Three parts are closely related and should be considered together.

From the product perspective, the material is the constituent element of the product (structure), and the construction process is the method of manufacturing the product. In other industries, product-material-process integrated design is nothing new, but in engineering, the concept of integrated design has not yet been fully realized. With the help of intelligent construction, it is imperative to advance this concept to promote the development of intelligent engineering design. Figure 6.10 shows the basic idea.

In Fig. 6.10, different structural layers (including roadbeds) have different performance requirements for the road structure system according to the input characteristics. The material properties in each structural layer need to be matched. To ensure that the structure achieves the required performance, it is necessary to select the constituent elements for each layer. The mixture needs to be formulated according to the material properties of each component, such as adding binders to enhance the strength. The construction process needs to be provided on this basis. This integrated design is a system design, and the knowledge must be placed in the database.

> The system's input is the driving load (temperature and humidity influence must be considered simultaneously). According to the law of continuum stress propagation, the additional stress caused by the driving load spreads from top to bottom, and the stress becomes smaller and smaller. Therefore, the structural performance requirements are gradually reduced from top to bottom, and the corresponding structure is also designed in layers.

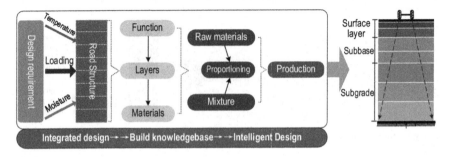

Fig. 6.10 Integrated design of roadway structure, material, and construction

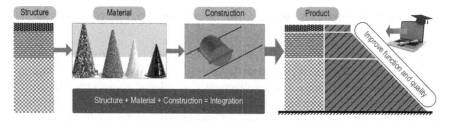

Fig. 6.11 Integrated design of roadway

The unified structure and materials design in road engineering has been achieved, but the combination with construction technology has not yet been fully achieved. Roadbed and track design are still separated in railway engineering without considering construction technology. In the design of airports and urban roads, the situation is similar. Without the participation of the construction process, no matter how well the structure and material are designed, the design may not be fully realized. There are only raw materials and proportions, but no cooking method exists. Therefore, the construction process is most lacking in the integrated design of roadways, including raw material preparation, mixture mixing, paving, and rolling. The rolling process is the key to the construction of the structure.

Figure 6.11 shows the framework of the integrated design using the neuro-expert system. The trained (learned) design system should provide several design schemes based on the design data and then optimize and make decisions. Life and cost can be used as goals of optimization to improve performance and quality.

The subbase and base construction process includes material preparation, mixing, paving, and compaction. It is noted that the current construction process (parameters) is imperfect for the construction process. There are not many parameters that can be adjusted depending on construction machinery. With the development of intelligent construction machinery (pavers, road rollers), those process parameters can be adjusted in a more extensive range, and there will be more choices. Of course, these are based on massive data. Advances in construction technology will also promote changes in design methods.

In addition, since the pavement surface characteristics and alignment are closely related to driving safety, the integrated design should also be combined with the geometric design of the roadway, promoting intelligent transportation development.

(4) **The trend of Intelligent engineering design**

Intelligent engineering design is still in its infancy (or even the conceptual stage), but its development prospects are generally optimistic. This is also an inevitable trend in the development of design technology. Much practice shows that the combination of theory and experience is the core of the design. Multidisciplinary knowledge is one way to improve design, and intelligent engineering design is bound to be inseparable.

1. **Integrated Intelligent design with knowledge as the core will be the mainstream development**

Regardless of the industry, any design issues will inevitably use existing experiences (historical data). These experiences are knowledge in AI. Therefore, at this stage and for some time in the future, the development of intelligent engineering design still needs to rely on the powerful functions of expert systems in knowledge processing and then use appropriate intelligent algorithms to make up for the limitation of the learning capabilities of expert systems.

The design work involves a wide range and many influencing factors for road or railway projects. Forming a regular and complete knowledge base like structural engineering is complicated. Therefore, the combination of expert systems and intelligent algorithms is essential. On the one hand, it is necessary to collect and compare successful engineering design examples and store them in the knowledge base so that existing knowledge can guide the design work. The implicit knowledge is obtained from the data to supplement and improve the deficiencies of the expert system.

2. **3D design and virtual reality technology will be widely used**

The three-dimensional virtual design originated from product surface modeling design in the 1960s, later developed into solid modeling design and virtual prototype technology, and combined with other software such as CAE (Computer-Aided Engineering), CAPP (Computer Aided Process Planning), CAM (Computer-Aided Manufacturing). It has been well integrated with VR (Virtual Reality) and AR (Augmented Reality) technologies and has developed into a design platform integrating information, networking, and intelligence. Three-dimensional design in the industry runs through the entire product design, manufacturing, and assembly process, significantly reducing the difficulty of collaborative design and rework in the manufacturing and assembly process. Three-dimensional design in the civil engineering industry should also follow this method.

For road and railway engineering, the development of unique three-dimensional design technology is promising, which needs to be combined with virtual reality technology and intelligent technology and can simulate the performance and function of the structure.

Three-dimensional design is like three-dimensional cutting; you can find many ill-considered problems in two-dimensional design. For geometry design, it can more intuitively and accurately express all the geometric features of the route, simulate the movement process of the vehicle from a kinematics perspective (virtual test/computer simulation test), and obtain valuable parameters for linear design. From the perspective of dynamics, it is possible to simulate the distribution of the structure's stress, strain, and fatigue under the driving load and analyze each structure layer's mechanical characteristics = , which are very useful for optimizing the structure and material design. In addition, one should also simulate the water and temperature effects on the structure's performance, drainage, and surface function. In short, the three-dimensional design of road and railway engineering, in addition to simulating the route's geometry and its relationship with traffic, is necessary to simulate the performance and function of the structure, which also helps improve and update the knowledge base.

3. The design platform based on IoT will be further developed

As a subsystem of an intelligent construction system, the intelligent engineering design system has a robust human–computer interaction function that is very necessary. Engineers must effectively intervene and work together in the design process, consider the connection with the construction phase, and provide a unified data model and data exchange interface. In principle, the above goals can be achieved with the help of a design platform. Still, the existing design platform is relatively independent, and data is challenging to share. Therefore, the Internet of Things technology is needed.

The Internet of Things is born with the popularization of networking. The design platform based on the Engineering Internet of Things is a part of the information platform of the intelligent construction system, and its architecture and data format standards are unified. Therefore, the design data running on it will be shared with the construction and maintenance subsystem so that the data can be shared.

The above contents take road engineering as an example to discuss the feasibility of intelligent engineering design. Based on the successful experience of other industries, it is feasible to use expert systems and machine learning, which applies to railways, airports, and urban roads. The foundation of intelligent engineering design has been established, and now it is time to encourage more people to participate.

6.3   The Rise of Intelligent Construction

Transforming design drawings into products is called "manufacturing" in the industry and "construction" in the engineering world. Although the titles are different, the meanings are the same. Intelligent manufacturing is the main battlefield for the industrial world, and intelligent production is the main line. Correspondingly, the main battlefield of intelligent construction in engineering is the construction site, and intelligent construction is the main direction.

Intelligent engineering design is still conceptual, but intelligent construction has already begun. The appearance of an "intelligent construction site" is an example. Although there is not much intelligence yet, the direction is correct. An information platform centered on informatization and networking has been formed, but intelligence needs to be developed. Promoting the intelligentization of various technologies during construction can improve the project's quality and efficiency needed to construct highways, railways, and airports.

In recent years, "smart construction site" as a fashionable vocabulary has appeared in the construction fields of highways, railways, municipalities, and airports. But in essence, most of them are just different opinions of the information and they are doing informationization and network management without wisdom. The core of "wisdom" is creativity, and its meaning

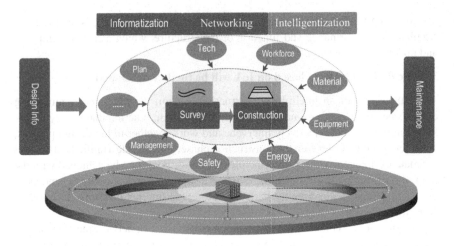

Fig. 6.12 Intelligentization of road construction

is higher than "intelligent". The term intelligent construction sites (should be derived from the "Smart Planet") proposed by IBM in 2008, but "Smart + " does not have the exact meaning of "intelligent or wisdom." There is still a long way to go from smart to wisdom.

Because track plates and rails in railway engineering are all prefabricated parts, their intelligence belongs to intelligent factories. The on-site is mainly automated assembly. Figure 6.12 shows the intelligence in the road construction process.

The components involved in the road construction phase are vibrant (Sect. 1.7). In principle, intelligence involves every aspect of the construction phase, but at this stage, we still need to focus on the key points, not everything. Building an intelligent construction information platform (construction part) needs to be considered first by placing the construction information on the platform, achieving informatization and network management, and selecting key parameters to achieve intelligence that needs to be implemented, as shown in Fig. 6.12.

An information platform based on the Internet of Things is essential. Under the coordination of the information platform, the design information from the design stage will be transmitted to the construction stage through the network. The construction management unit distributes the tasks to the construction units through the platform and then begins to provide plans, construction organization design, construction of structures, and project management.

Various popular information platforms have realized network-level information management, and there is still a gap in intelligentization. With the foundation laid, there is only one step towards intelligentization, but it is not easy to take this step. It requires multi-disciplinary knowledge and sufficient data as support.

For the convenience of description, the construction shown in Fig. 6.12 is divided into two parts: structural construction (construction stakeout and layered construction) and service guarantee (plan, workforce, materials, equipment, management, etc.). The following is a brief explanation based on the technical approach of "perception, analysis, decision-making, and execution."

(1) Service guarantee system

Service guarantee is to manage people, finances, and materials well to support construction. Its composite components in various areas characterize this part. Currently, there is no general intelligent system to do this. It must be divided into several subsystems or modules according to the specific situation and realized by different intelligent technologies.

1. Preparation of plans

During the construction phase, various plans need to be prepared. Although the content is different, the rule is easy to follow, and the format is relatively fixed. This feature determines the feasibility of adopting an expert system, converting various plans into "rules" to form a unique intelligent system. The following takes the construction organization design as an example for explanation.

The construction organization design is to formulate economic and reasonable construction plans according to relevant technical policies, construction project requirements, and construction organization principles to ensure that the construction tasks are completed according to the established goals.

According to the above statement, technical policies, construction project requirements, and construction organization principles are the primary basis for design plans, which are relatively fixed. You can use tools, such as IF–THEN and ANN, to convert them into various rules for construction to establish an intelligent system. When the project information is input into the intelligent system, it will automatically optimize the target and generate an economical and reasonable construction plan (it can be understood as a decision-making process), as shown in Fig. 6.13.

The construction organization plan is about how to ensure the construction. An expert system can be used, but some parts can also use other intelligent technologies. For example, for the construction progress, ANN can predict, and the training data

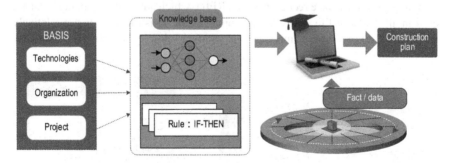

Fig. 6.13 Intelligent preparation of construction organization and plan

is the past successful experience. When designing the input of ANN, many factors must be considered, including the amount of construction, the number of personnel, the number of equipment, material supply, capital guarantee, etc., and they must be converted into digital quantities.

2. Personnel safety management

The essence of people safety management is the control of human behavior. This is a very complex problem that can be included in the category of social sciences, and it is difficult to completely solve the problem by simply using natural science methods. For example, traffic control, essentially the control of driver behavior, cannot be solved by using some mathematical formulas, which is also one of the fundamental reasons for the slow progress of current intelligent transportation research.

According to the technical requirements of control (see other volumes of the series), the prerequisite for quantitative control of an object (person or object) is to have a mathematical model of the controlled object. Therefore, in all people management and control problems, the focus is on how to model people's behavior. At present, it is impossible to establish a human behavior model in engineering construction, and it is mainly based on qualitative management and control.

The management and control of people in the construction phase mainly involve three aspects—identity, location and behavior. These three aspects of information need to be acquired by sensing technology, among which the acquisition of human identity information adopts technologies such as face recognition and RFID, and the human location information adopts satellite positioning technology (mainly mobile phone positioning), and human behavior information is relatively complex. With diversity, computer vision technology can currently be used to obtain action information.

In principle, after sensing the relevant information of the construction site personnel, it is transmitted to the information platform through a wireless network (such as 5G) for analysis and decision-making. Then a feedback control command is issued, as shown in Fig. 6.14.

The current intelligent management of personnel is relatively simple, mainly using technologies such as face recognition and RFID, which are relatively mature and are used in all walks of life. Since the personnel at the construction site are relatively fixed, this kind of simple management is relatively simple, and it is relatively easy

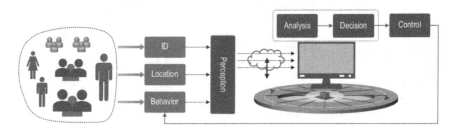

Fig. 6.14 Personnel safety management

to use some intelligent technologies for management. You can refer to the practices of other industries.

3. Management of materials and equipment

Materials and equipment are the basis of the construction stage, and they all belong to "objects." Their management issues have standard features, so they are introduced together. The materials mentioned here mainly refer to various soil, aggregate, asphalt, cement, and steel bars used in filling projects. The equipment includes various engineering machinery, testing instruments, and mixing stations.

When it comes to objects, IoT is usually mentioned. As long as the item is installed on the corresponding sensing terminal and connected to the network, every item's movement will be monitored and managed easier. However, the current implementation of interconnection and intercommunication will involve two problems. One is the cost issue, which is a heavy burden on enterprises; the other is that suitable sensing devices may not be found and may need to be developed separately.

Material management is an important task, and the management method has developed from manual operation to the computer-based informatization stage. The source, purchase, quality, acceptance, quantity, and storage of materials must be managed and controlled in a unified manner. The RFID tag can be configured according to the batch and type. The origin, specifications, and related parameters can be written into the RFID chip to realize the inbound and outbound management. As for conducting intelligent management, it is believed to be not difficult to do it with the previous foundation.

Location information should also be known for equipment management and identification, especially for construction machinery and testing equipment. This type of configuration is widespread, and the management is relatively simple. The intelligentization of equipment performance and functions should be considered (Fig. 6.15). It is still under development with a promising future.

There are two aspects of the intelligence of construction machinery. One is the intelligence of performance, and the other is the intelligence of functions. The intelligentization of performance belongs to intelligent manufacturing, mainly manifested in the ability to automatically perceive the machine's operating state, learn and master

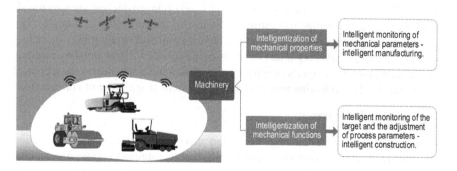

Fig. 6.15 Intelligent properties and functions of construction machinery

diagnostic skills, and diagnose and maintain (or information feedback) failures. For construction, the concern is the intelligentization of its functions, which is most closely related to the construction quality and efficiency. The critical performance is to automatically perceive relevant information, control the machine, and provide proper responses based on the interaction between machinery and road materials.

4. **Safety and management**

Construction safety issues have always been paid attention to, and the management departments have precise requirements and have designated relevant rules and regulations. What needs to be done is to use intelligent means to implement it. You can refer to the following steps to the intelligentization of rules and regulations.

The first step is to transform rules and regulations into knowledge in the form of "rules" used by expert systems and ANNs. The second step is to use appropriate sensing terminals to perceive relevant information related to construction safety in real-time. Finally, the trained and intelligent model is used to process the sensed information in real-time, analyze and make decisions to issue execution instructions.

In addition, other machine learning algorithms can also be used to deal with the problem. For example, the problem of on-site personnel wearing helmets is a simple safety monitoring problem. The drone can be used for tracking after taking images. Then the trained deep learning model can be used for analysis to identify the number of people who are not wearing helmets and provide a basis for decision-making and management (execution). Intelligent means need to be used to discover and solve this problem.

Management runs through the entire process, from accepting construction tasks to project acceptance. At present, information management has been realized, and it has realized intelligence. Comprehensive intelligent management should be based on an information platform with the Internet of Things as the core. Different intelligent technologies should be adopted for different objects and needs. Management involves every process in the construction process and does not exist independently. These will be discussed further later.

(2) **Construction of the structure**

The structure mainly refers to the pavement structures of roads, railways, and airports. The auxiliary facilities, such as retaining walls and drainage ditches, are not considered (there are also intelligent construction issues, but they are relatively simple). The geometric alignment is conducted on the construction site (staking out) according to the design documents regarding the construction process. Then suitable materials are selected, and each layer's construction is carried out according to the technical requirements. The following introduces different steps in the order of construction.

1. **Construction stakeout**

Construction stakeout is "moving" the design plan on the drawing to the construction site. Due to the total station, the traditional manual measurement is avoided, and the efficiency is improved. The new total station (commonly known as the measuring

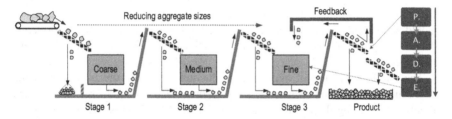

Fig. 6.16 Aggregate production process and intelligent process

robot) has realized the automation of construction lofting, and its efficiency is about 6–7 times that of the traditional method.

Regarding the intelligentization of the stakeout, automation is misunderstood as intelligentization. Stakeout is a measurement process without much intelligence, but automation is enough. Intelligence is mainly embodied in measuring instruments, such as the intelligence of the total station's performance (similar to the intelligence of the sensor) or an autonomous correction of measurement errors. For details, please refer to the technical manual of the total station.

2. **Material production**

Material production is mainly for aggregate. Aggregate is the primary raw material of various mixtures, which is critical for the quality of the structure. Aggregates are produced by crushing-screening combined equipment, including fixed and mobile types. To find the process to apply intelligence, it is necessary to understand the production process.

Figure 6.16 shows the production process from aggregate to the finished aggregate product. The crushing equipment breaks the aggregate from large to small aggregate particles by mechanical actions. Then screening equipment separates the aggregates of different sizes and meets the gradation requirements. After analysis, it was switched to intelligent control in the third stage (crushing to produce fine aggregates). The key is the perception of aggregate information (size and shape).

The perception of aggregate size (including quantity) is relatively simple, and the screening machine can realize it. The aggregate shape is expressed by the shape factors (long axis/short axis). Digital image technology can measure the aggregate's long axis and short axis.

The perceived aggregate information will be input into the intelligent model and compared and analyzed with the standard value. The decision can be that it needs to be broken again, or the process parameters need to be adjusted. Finally, feedback control will be conducted through machinery control.

The standard value can be set according to the production target (5–10 mm). The aggregate shape is preferably a tetrahedron or a cube, and the shape factor is preferably close to 1, such as 1.2.

Regarding the intelligent model, the training data should include the process parameters of the equipment in addition to the aggregate information (size, shape, crushing rate, etc.). These data are relatively easy to obtain, and it is easy to determine the control requirements. In addition, when finished products leave the factory, they should be marked with electronic tags (RFID) in batches, and relevant information should be input as a basis for traceability, which is convenient for management and quality control.

The intelligentization of aggregate production is mainly the intelligentization of production equipment, which is rarely reported. This is a valuable and exciting direction (including the crushing method) worthy of in-depth study and is also of concern to the engineering community.

3. Material mixing

Material mixing is mixing and stirring various materials to produce a mixture according to the design requirements. It is divided into the roadbed, base, and surface materials. Although the materials, mixing equipment, and processes are not entirely identical, the basic principles are similar. The ideas for achieving an intelligent mixing process are the same, emphasizing the measurement (sensing) and control of the mixture's components.

Figure 6.17 shows the mixing equipment (mixing stations) of stabilized soil, cement stabilized gravel, asphalt mixture, and cement concrete. Mixing the base and surface materials is more complicated and focuses on intelligence.

The control system is the key to improving the mixing quality in the material mixing process, and automation has been achieved. It is not difficult to realize intelligent control on this basis, as long as it is implemented by perception, analysis, decision-making, and execution (control). Figure 6.18 shows the keys to the intelligent mixing process, in which the intelligent model needs to be trained in advance to learn the material design data.

Regardless of the type of material mixing, although the specific process is not entirely consistent, strict control of the proportion of each component is common, and it is also the key to intelligent control. The current automated mixing has been done very well. Intelligence is not the most urgent issue, and the top priority is the information sharing of the mixing station. Most construction companies are unwilling

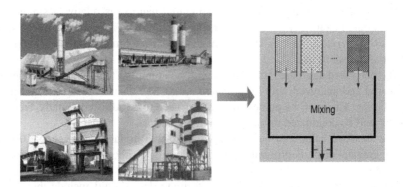

Fig. 6.17 Mixing equipment

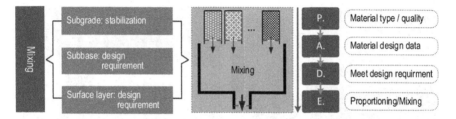

Fig. 6.18 Keys to applying intelligent material mixing

to provide the production data of the mixing plant, which affects the comprehensive analysis of construction quality. Breaking the phenomenon of data islands is a problem that needs to be solved urgently, and it is the same in the entire intelligent construction.

4. Material transportation

Transportation is transporting materials from the reclaiming yards or mixing plants to the construction site. The main task of this process is to monitor and manage the identity (RFID perception) and location (satellite positioning) of materials and vehicles (items) (similar to the unified dispatch of taxis). In addition, temperature sensors are also required to monitor the temperature for the transportation of asphalt mixtures. The current monitoring system based on the information platform has achieved the above goals.

With the full deployment of intelligent construction, it is also necessary to establish the interconnection between vehicles-mixing stations and equipment (graders, pavers, road rollers, etc.) to realize the Internet of Things to facilitate unified command collaborative work.

5. Paving

The materials' production, mixing, and transportation are prepared to construct structures. The construction stage starts with paving. The first thing that needs to be controlled in the paving process is the layer thickness and surface smoothness. The laser leveling system equipped on the paving machine can automatically control these. The other thing that needs to be controlled is the segregation of asphalt mixture. It is affected by material design, mixing, transportation, loading and unloading, and paving, which cannot be fundamentally solved. After adopting intelligent control technology, the segregation problem can be improved.

Segregation refers to the phenomenon that the different components in the mixture cannot be well mixed. Segregation is a kind of variation in structure, which will cause uneven distribution of the structure's internal composition and physical and mechanical properties, which is one of the main reasons for premature pavement damage. A typical segregation phenomenon occurs in the paving of asphalt mixtures, including material and temperature segregation. The temperature distribution on the paving surface is quite nonuniform, as shown in Fig. 6.19.

Fig. 6.19 Material and temperature segregation of asphalt mixture

Fig. 6.20 Transfer of material segregation

A lot of practice shows that there will be segregation transfer problems in the mixing, loading, transportation, and unloading process. This segregation will be further amplified during the paving process, separating fine aggregates and between asphalt and aggregates. Figure 6.20 illustrates the transfer process of segregation, and controlling segregation should also start from these stages, including material design.

Regarding temperature segregation, it should be closely related to material segregation. At present, thermal infrared sensors can be used in the paving process to automatically sense the temperature distribution of the paving surface and monitor the temperature states in real-time, as shown in Fig. 6.21. But how to control temperature segregation still needs to be studied.

Segregation does not only occur in asphalt mixtures. In all multi-component mixtures, a certain degree of segregation occurs. It is almost impossible to avoid

Fig. 6.21 Identification of temperature segregation

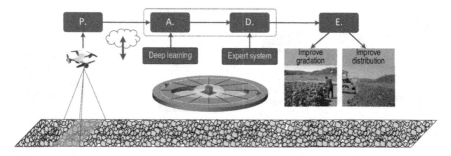

Fig. 6.22 Realization of intelligent construction of the granular base

it, and the key is to control it within the allowable range. The definition of this allowable range is more complicated. It involves the impact of variation of structure performance on the superstructure and driving safety, which is beyond the scope of this book. Let's start with the roadbed and make a simple analysis of the intelligent problem of the paving process.

Due to insufficient attention, the roadbed is often treated as an earthwork project. The "nearest retrieving" method is applied in the field, making the sources and types of fillers more complicated to control accurately.

In addition to uneven distribution (segregation) of coarse and fine materials, there is also a common problem of unreasonable gradation. If these problems are not solved during the paving process, they will be more challenging during the rolling stage. If there is no quantitative requirement for grading and particles are not sieved, intelligent technology is feasible, as shown in Fig. 6.22.

In Fig. 6.21, the constructed granular base is photographed (perceived) by a drone and then sent to an intelligent model (such as deep learning) to analyze the size and distribution of aggregate particles. Then, the data is sent to the expert system to compare with experience and knowledge to improve grading and distribution, issuing instructions for execution. Other parts are run on the information platform except for the perception part.

Regarding execution, it includes two parts: improving the gradation and particle distribution. The detailed application includes improving the gradation by adding particles of different sizes, using mechanical or manual methods to process coarse particles, and guiding construction machinery to improve the distribution of fillers on the paving surface.

Roadbed construction machinery includes bulldozers, scrapers, graders, excavators, loaders, and road rollers. The intelligentization of construction machinery mainly involves the interaction between machinery and soil. In addition to the intelligentization of road rollers, other construction machinery mainly uses high-precision satellite positioning, which can achieve informatization and partial networking. However, complete interconnection and intercommunication cannot be achieved, as shown in Fig. 6.23.

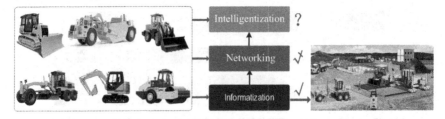

Fig. 6.23 Technical status of subgrade construction machinery

At present, automatic machine-guided construction (Fig. 6.22) is the most popular, which can achieve precise work on soil and aggregate, making the construction from mechanization to informationization and automation. If it is networked, it can reach the level of interconnection and will play a more significant role under the unified information platform.

Automatic Machine Guidance (AMG) construction has developed rapidly, mainly using high-precision satellite positioning and various angle sensors, laser sensors, etc., to realize the informatization and automated construction of some construction machinery. For example, in the excavation section, through the guidance system, the bucket position and excavation parameters of the excavator can be displayed on the screen in real-time to realize the guidance of the elevation and slope; the leveling system on the grader can realize the automatic control of the slope and elevation. Automation technologies have developed rapidly. Although they do not have much intelligence, they have solved some practical problems in construction and have already met the needs. There is no need to pursue intelligence too much.

The intelligent paving problem of the roadbed has been discussed above. The paving of the base layer and the surface layer have commonalities. The following uses asphalt paving to illustrate the key points to achieving intelligence.

As mentioned earlier, controlling paving thickness and smoothness has been achieved with the automatic leveling system of the paper, which has now met the needs at construction sites. Because the rolling of asphalt mixture is greatly affected by temperature, the higher the initial density after paving, the better the construction quality. The initial density is related to the paver's process parameters and the screed's vibration characteristics. Therefore, this part is suitable for intelligent technology.

Many years ago, in the paving of the asphalt surface layer, we did a lot of experimental research on the relationship between the paver process parameters and the initial compaction degree. It is found that different process parameter combinations have a more significant impact on the initial density, and a good process combination can increase the initial density by more than 5%. From the perspective of the vibration response of the screed, the process with high vibration intensity does not necessarily have a high density. When the initial density is high, the vibration of the screed has specific characteristics, which will be reflected in the time and frequency domains. People will have different processing methods to extract this feature or control index.

The intelligence in the paving process, starting from the vibration characteristics of the paver screed and the basic requirements of intelligent construction, is shown in Fig. 6.24. This is an intelligent solution based on empirical data.

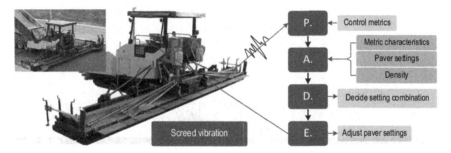

Fig. 6.24 Intelligent points of the paving process

In Fig. 6.24, the vibration signal of the screed is recorded by sensors. Extracting vibration characteristics (control indicators) is complex and requires professional knowledge and test data. The intelligent control system needs to analyze the vibration characteristics of the screed, the corresponding process parameter combination, initial density, etc., which are required for model training. The decision is based on the principle of achieving maximum density using the corresponding process combination. Control (execution) is the action taken by adjusting process parameters. If it is an intelligent paver, it should also be able to adjust parameters automatically.

The above intelligent scheme is elementary, and the control index is also based on experience. In addition, the interaction between the paver screed and the paving layer can also be analyzed, and better control indicators can be theoretically obtained to propose a better solution. This is a good research direction, but it is more complicated.

The intelligence of the paver has both process classification and density prediction problems. Different intelligent algorithms or combinations, such as ANN, support vector machine (SVM), and expert systems, can be selected. Without a large amount of experimental data, you can use reinforcement learning to learn and optimize the paving process.

During the development of intelligent paving, from raw materials to mixing, transportation, paving, and rolling of asphalt mixtures, the interconnection of various items, including construction machinery, is also emerging (Fig. 6.25), promoting the development of the Internet of Things. For example, segregation and temperature distribution during the paving process can guide the subsequent compaction.

In summary, the intelligentization of the paving process of asphalt mixtures is to improve the uneven (segregation) distribution of materials and increase the initial density. Other aspects are mainly informatization, which has met the current construction needs. The focus of intelligent construction is the compaction stage.

6. **Rolling**

For filling engineering projects, rolling is the last process to turn the filler into a structure, and it is also the most critical construction process. Its intelligence level is currently the highest, representing the overall level of intelligent construction. Construction machinery involves various road rollers, and intelligent control technology is intelligent compaction. In addition to applying intelligent compaction in

Fig. 6.25 Interconnection throughout the construction process

asphalt mixtures, the application in other construction materials is mature. A more detailed discussion can be found in the related publications.

Rolling is the interaction between the road roller and compacted material, and it is also the process of changing the material from the loose state to the compacted state. Due to the different filling characteristics of each layer, the utilized rollers are different. Single-drum vibratory rollers mainly compact the roadbed and base layer. The asphalt surface layer is mainly compacted by double-drum vibratory rollers, including the combination with rubber-wheeled rollers.

The practice has shown that different rolling fillers' corresponding intelligent and application methods are not the same. The rolling method of asphalt mixture is the most complicated. Construction machinery should have the ability of interconnection and high-precision positioning, which is the embodiment of information and network, as shown in Fig. 6.26.

In the rolling process, the most critical work is compaction quality control, which is also the focus of intelligence. The traditional control method is "point-type" sampling and testing, and the control indicators include density (physical indicators) and modulus (mechanical indicators). The density test includes the sand filling, core, and non-destructive methods (such as a non-nuclear densitometer). Modulus tests can be plate load tests, falling weight deflectometer (FWD), or lightweight deflectometer (LWD).

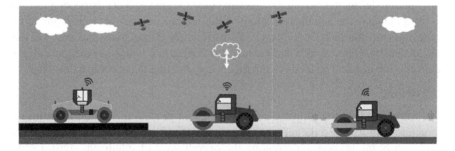

Fig. 6.26 Capabilities of a roller

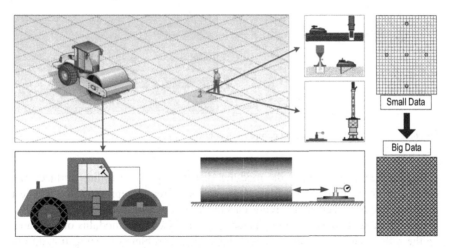

Fig. 6.27 Comparison of point control and intelligent compaction control

The main drawback of traditional control methods is that they cannot conduct comprehensive and real-time inspections on the rolled surface, and sampling control points may not be sufficiently representative. Intelligent compaction control methods make up for these shortcomings. Still, intelligent compaction is similar to a moving plate load test and cannot directly detect physical indicators (density). Therefore, the density needs to be tested after intelligent compaction, and the mechanical index cannot replace the density.

Modulus and density are two independent performance indicators (physical property parameters) that are statistically relevant. The modulus of the compacted body in the compacted state is much larger than that in the loose state.

Figure 6.27 compares point control and intelligent compaction control. It can be seen that, compared to point control, intelligent compaction control covers all points on the rolling surface, similar to the relationship between "small data" and "big data." Some changes might happen to the evaluation and control systems.

Intelligent compaction adopts continuous detection, covering the entire rolling surface and the entire rolling process. The obtained data is vibrant, and various analyses can be performed.

For example, according to the roller response information obtained by continuous detection (perception), the relationship between the compaction modulus (or other mechanical indicators) and the number of rolling passes can be obtained in a test section. The modulus of each point on the rolling surface can be generated. Figure 6.28 shows that through the distribution map, a lot of valuable information and knowledge can be mined from the map.

Figure 6.28 (left) shows the modulus change at a certain point in the green area D with rolling passes. The following information can be mined: the best rolling pass $N_D = 7$ in the D area; the D area has reached the solid compression in a stable state; packing in the D zone has good compactability.

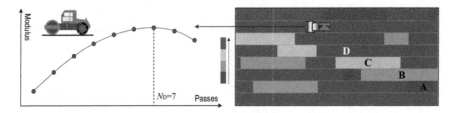

Fig. 6.28   An example of intelligent compaction control

Figure 6.28 (right) shows the modulus distribution at the rolling surface. The change range is divided into four areas. The red area A represents the smallest modulus, and the green area D represents the largest modulus, $E_A < E_B < E_C < E_D$. Under the same number of rolling passes (seven passes), the modulus of each area on the rolling surface is quite different, indicating a problem of segregation during paving. The problem of unevenness (the unevenness of the underlying layer will also have an influence, but there are ways to separate them). The intelligent compaction control can play a role according to the real-time distribution of the compaction state by guiding the roller to perform additional compaction in low-value areas (the optimal number of compaction passes for each area is different). This will improve the distribution of the properties (modulus) of the compaction and control the modulus difference $(E_D - E_A)$ within the allowable range. This is not possible to implement with point control. If an intelligent roller is used, this process will be completed automatically.

The right picture of Fig. 6.28 also reveals the segregation problem's consequences: the compacted subject's performance may vary. Intelligent compaction turns the segregation problem from a qualitative description of the paving stage to a quantitative analysis. It determines the size of the difference, which provides data support for controlling (improving) the variability. You can also use the information platform to trace the various paving, mixing, and raw material production processes to find the cause of segregation and its solutions.

In addition to intelligent compaction, another method is widely used to control compaction quality by recording the number of rolling passes. This is the reproduction of the traditional rolling pass control method, but it has changed from manual recording to automatic recording in the past. Generally, high-precision positioning is used to achieve this process, as shown in Fig. 6.29.

This empirical control method requires that the test and construction sections have the same underlying layer, packing, roller, and compaction process parameters. The packing is uniform, as shown in Fig. 6.29. If one of them is not satisfied, this control method is unsuitable. The non-uniform compaction characteristics, as shown in Fig. 6.28, are unsuitable for this method.

Most rolling asphalt mixtures currently use the rolling pass control method, which is not intelligent, and the effect needs further observation. In addition, temperature management is also critical in rolling, providing a basis for intelligent control of rolling progress.

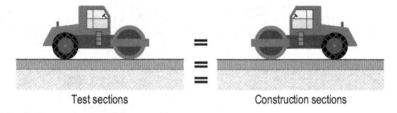

Fig. 6.29 Conditions of the rolling pass control method

Restricted by the characteristics of the asphalt mixture, the rolling operation must be completed at a temperature higher than 80. Otherwise, it will be difficult to compact. When the paving is completed, the biggest concern is how long it will take to drop to 80 °C to control roll. Intelligent technology can be used to make predictions.

Equipped with an intelligent control system on the roller, it can intelligently predict the remaining time (t) when it drops to 80 °C based on the perceived surface temperature. The intelligent system must be trained to master the predictive skills before working.

Figure 6.30 shows the training method using ANN. The input training data is the surface temperature of the paving layer, the ambient temperature, the paving layer's thickness, and the lower layer's temperature. The output is the remaining time when it drops to 80 °C. When working, it is necessary to input the thickness of the paving layer, the ambient temperature, and the lower layer's temperature into the intelligent system. The system senses the surface temperature of the paving layer in real-time through the temperature sensors. After analysis and decision-making, it predicts the remaining time when it drops to 80 °C. According to this, the rolling progress is controlled.

Intelligent compaction continuously perceives the response information of the roller during the compaction process, obtains the mechanical performance data of the compaction through complex analysis, conducts analysis and decision-making, and finally adopts feedback control measures to ensure the quality of compaction.

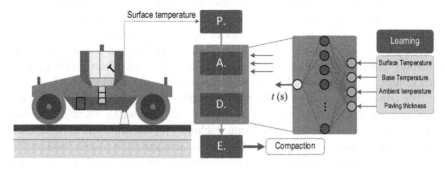

Fig. 6.30 Intelligent control of the rolling progress of asphalt mixture

Fig. 6.31 Assembly-type component fabrication and construction

This is completed by the intelligent compaction control system assembled on the road roller, which fully embodies the processes of "perception, analysis, decision-making, and execution" of intelligent construction. It is a successful engineering example at present. See the following discussion for details.

7. **Manufacturing and installation of fabricated structures**

The above is the filling structure formed by paving and rolling on site. Another type of structure in highway and railway engineering is processed, manufactured in factories, and installed on-site, called a prefabricated structure. For example, some concrete pavements of highways, railway tracks, bridges, and stations are all prefabricated structures, as shown in Fig. 6.31.

The development history of the fabricated structure is only a few decades, and the production of industrial products inspires it. It is hoped that various components can be manufactured in the factory and then assembled on the construction site to form a structure. In the 1960s, this idea was first realized in construction engineering and later extended to the road and railway fields. Its benefits include fast construction speed, low production cost, guaranteed quality, and easy standardized operation (standardized design, factorization manufacturing, and mechanized installation).

Assembled components are produced in factories, and the intelligence of the production process is one of the categories of intelligent manufacturing currently developing rapidly. On-site assembly is mainly mechanized. For example, the sleepers and rails of ballasted railway tracks and the track plates and ballastless tracks have been mechanized installation operations, as shown in Fig. 6.32.

For component installation, the most important thing is to control the installation accuracy (the same is valid for assembling other products). Therefore, it is an excellent choice to guide the construction machinery for construction and installation through a machine (such as robots), which can be automated. Of course, based on the understanding of the characteristics of intelligent construction, the lofting robot (total station) "perceives" the size information during the component installation process,

Fig. 6.32 Field installation of prefabricated components in railway

"analyzes" the difference with the design requirements, and then makes a "decision" on whether to adjust the position of the component. Finally, it "executes" the instruction to control the accuracy. This part of the intelligence is relatively simple, and it is not impossible to regard it as a higher level of automation.

(3) **On-site laboratory**

The experiments done in the laboratory, whether in operation or data processing, are standardized operations carried out according to the technical requirements and do not require too much intelligent processing. With the popularization of information technology, many test instruments are equipped with computer control operation, informatization, and automation. It is also easy to develop networking on this basis. It is possible to realize remote real-time monitoring of the laboratory's test process and remoted management using the information platform.

(4) **Inpsection robots (under development)**

An on-site inspection is essential during the construction phase and has precise requirements in the technical standards. The required testing equipment, such as density meter, plate load tester, FWD, LWD, etc., is also relatively common. Most of these devices are controlled and operated by very convenient computers. Some have reached the level of automation and can meet the needs of engineering construction. The following introduces the inspection robot that is under development.

A robot is a mechanical device that can automatically conduct specific tasks (such as inspection and processing) through remote control or programs, relying on its power and control capabilities. At present, inspection robots are becoming a hot topic in many fields. Some products have also appeared in transportation, such as track inspection robots, road inspection robots, and drone inspections. One of the main features of these products is upgrading traditional testing instruments. The testing principles and methods have not changed substantially, but they have become more automated. Real-time data transmission can also be carried out through the

Fig. 6.33 Principle of multi-purpose inspection robot

network, often used during operation. The inspection of infrastructure performance and functions provides a basis for maintenance decision-making.

Although the traditional point detection is accurate form filling projects, it is time-consuming and labor-intensive, and its representativeness is also problematic. However, the continuous detection performed by intelligent compaction is inconsistent due to the loading size (roller tonnage) and method (vibration process parameters). Different rollers and vibration processes produce different test results. Therefore, they are mainly suitable for the quality control of the rolling process. If the two aspects are combined to take advantage of their respective advantages, integrated innovation with technical principles to detection methods is a new idea for developing inspection technology. The multi-functional and multi-purpose detection robot introduced below is an integrated technology suitable for quality control in the construction phase. Its working principle is shown in Fig. 6.33.

Figure 6.33 shows the integration of continuous detection and steady-state vibration, which are coupling problems between vibration mechanics and wave dynamics. The specific difference lies in the boundary conditions. One is the interaction between the steel wheel and the continuum. The other is the interaction between the load-bearing plate and the continuum. Figure 6.34 shows some common uses of this inspection robot.

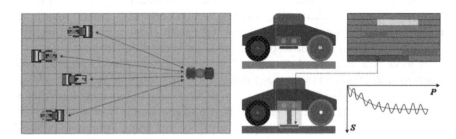

Fig. 6.34 Common uses of inspection robots

1. Used as small standardized continuous testing equipment

In the case of multiple rollers working together, the inspection robot can direct and coordinate the rolling operations of other rollers based on the continuous inspection results. If other rollers are equipped with an intelligent compaction control system, the inspection robot can also calibrate the intelligent compaction results of other rollers for normalization.

2. Used as combined equipment for continuous detection and fixed-point detection

Continuous testing is carried out in the designated area, and the distribution of different mechanical performance states is automatically divided according to the test results. The steady-state vibration test is carried out on different areas (the red low-value area in the figure). The plastic deformation, deformation modulus, and elastic modulus can be obtained simultaneously, which can be used to analyze the mechanical properties of compaction. In addition, the diameter of the carrier plate can also be changed to adapt to materials with different particle sizes.

3. Used as small accelerated loading test equipment

The steady-state vibration test system in the detection robot realizes the accelerated loading test of the compaction by changing the vibration frequency and the excitation force and is used to detect the fatigue characteristics of the compaction.

The "intelligence" of detection robots is still embodied in perception, analysis, decision-making, and execution. The remote control can also be carried out on the information platform, bringing convenience to construction quality control with good development prospects.

(5) Construction robot

Many people immediately think of various supermen in science fiction movies about robots. Most construction robots do not have a "human shape" but are equipped with a mechanical device with automated functions, as shown in Fig. 6.35.

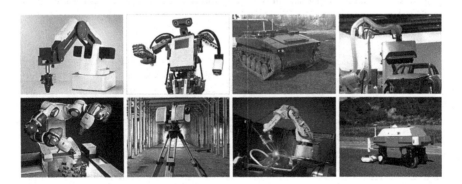

Fig. 6.35 Various robots for construction applications

Robots are divided into intelligent and non-intelligent types. Among them, non-intelligent robots are more advanced automation equipment to complete specific tasks. Intelligent robots need to be equipped with more sensors and intelligent algorithms with intelligent characteristics.

Construction robots are mechanical devices that automatically complete construction tasks. At present, they are considered unmanned construction machinery. Robots can be seen everywhere in the industry, and robots do many tasks like humans, such as welding and clean robots. Construction robots are on the rise in the engineering field, and all kinds of construction machinery previously operated by drivers will be operated by automation. Although non-intelligence is the primary method, automation will be further improved. When robots are applied for construction, safety issues require special attention, and there are currently no corresponding rules and regulations.

(6) **Deploying sensors to make the intelligent structure**

Intelligent construction aims to improve the quality and efficiency of construction and ultimately realize intelligent transportation infrastructure. While improving the quality, it can be made intelligent to have the self-sensing ability, constructing an informalized road with primary characteristics of intelligence. How to do this? It can be applied by deploying various sensors in the structure to make it perceive changes in the structure under the combination of traffic loading and environment (water, temperature). It integrates information to analyze, make decisions, and execute the decision, as shown in Fig. 6.36.

One of the purposes of perception is to carry out long-term tracking and monitoring of the structure's performance and grasp the evolutionary law. Another purpose is to provide necessary data support for intelligent transportation (including autonomous driving). Therefore, some sensors must be placed at specific structures to monitor deformation, stress, strain, humidity, and temperature during the construction phase.

Other data are obtained through sensors placed on the road surface regarding the data required for intelligent transportation and the above performance. It has developed rapidly as part of the Internet of Things (vehicles) and another information platform. The detailed discussion will be ignored since the relevant information can be found anywhere.

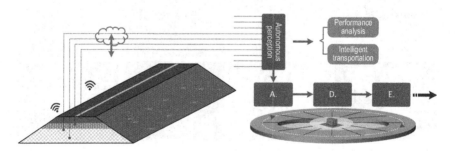

Fig. 6.36 "Perception" in transportation infrastructure

The above introduced some intelligent issues in the construction stage, but not all. Many aspects (including concepts and terminology) are still developing and changing, but the fundamental scientific principles will not change and needs to be learned. The era of intelligent construction is supposed to be welcomed with an open-mind attitude.

After completing construction, various data will be collected on the information platform. This is a batch of precious engineering data, including the initial state set during the operation period. At the same time, the sensing devices deployed at some critical locations will monitor the changes in the structure performance for a long time, and the relevant data during the construction period can be mutually verified. For AI, this is also excellent training data, laying the necessary foundation for intelligent maintenance.

6.4 Intelligent Maintenance has Started

After the facility's construction is completed, it becomes a "product" for vehicles (trains, cars, airplanes) to run on it, thus entering the operation period, accompanied by maintenance problems. The intelligence during operation is part of the Intelligent Transportation System (ITS). Nevertheless, an integrated and complex system should consider design, construction, operation, and maintenance, a research direction of intelligence. This section will elaborate on the intelligent maintenance issues.

Maintenance is a common problem for any kind of product during its use. The same problem exists during transportation infrastructure operation, and how to solve it is the task of maintenance.

The so-called maintenance is to decide whether maintenance is needed and the maintenance scale through a comprehensive analysis of relevant data and information and then put it into practice (i.e., execute). Therefore, as long as AI is used to complete these processes, intelligence can be achieved. The principle is not complicated, and the key lies in operating.

According to intelligent construction's essential characteristics, intelligent maintenance's focus has two aspects. One is the "perception" of information, which involves many detection technologies; the second is "analysis and decision-making," which AI needs to do. Appropriate algorithms and model training must be selected in advance. "Action" refers to maintenance and construction, introduced earlier.

(1) **Information sources and perceptions in the maintenance phase**

The information in the maintenance phase is the most critical basis for decision-making. Information is obtained through perception (detection) technology. Different methods obtain different information.

1. **Information sources for maintenance**

The maintenance-related information mainly comes from three sources, the one transmitted at the end of the construction, the one acquired by sensors embedded in the

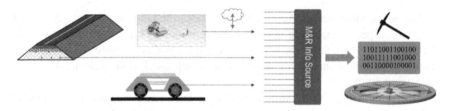

Fig. 6.37 Information sources and data mining of maintenance

structure, and the one detected on-site during the maintenance phase. This information will be input into the information platform and form an extensive database. A lot of knowledge can be mined from this information. This work requires a wealth of professional knowledge to be sufficient, and it also involves information fusion technology. Therefore, it provides a basis for making maintenance decisions, as shown in Fig. 6.37.

Among the three sources of information, the first two aspects of information are not yet complete. Currently, they mainly rely on on-site testing to obtain information. Therefore, the information derived from the on-site inspection focuses on the operation and maintenance phases, and many perception technologies are required. This is a rich and colorful field, including innovation in perception methods and equipment, and intelligence is a means to improve technology.

2. **What needs to be tested on-site**

During the operation period, regular inspection of existing structures is essential. The inspection mainly includes the structure's appearance and internal and performance characteristics, as shown in Fig. 6.38.

The shape feature refers to the geometric form of the structure. The variation of the geometry and surface characteristics will directly affect the smooth running of the vehicle, and accidents may occur in severe cases. Railroad engineering involves a lot of shape monitoring, regarding rails, gauge, height, curvature, twist, orbit, and

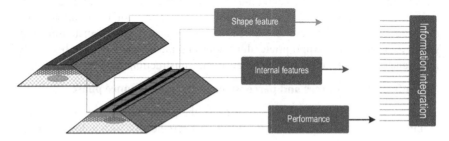

Fig. 6.38 Test contents of the structure

shape changes of sleepers, track plates, and fasteners. For roadway engineering, the shape features refer to the smoothness of the road surface, cracks, ruts, potholes, etc.

Internal features refer to defects inside the structure, such as cavities or voids, subsidence, etc. These defects will significantly impact the superstructure, such as track and road surface.

Performance refers to the physical and mechanical properties of the structure. There are many indicators to characterize the performance. Mechanical properties can be attributed to stiffness, strength, and stability, while physical properties depend on specific conditions. The physical properties of steel rails, asphalt surface layers, and concrete slabs have different indicators. The decreasing of structural performance indicators means that the maintenance time is approaching.

Variations in the structure geometry are related to the surface structure itself, but internal defects and performance variations in the structure will also affect it. For example, track and road surface irregularities and cracks are closely related to the uniform deformation of subgrade structures. When making decisions on maintenance, this information should be considered together.

3. **In-situ inspection technology**

In-situ inspection during operation is generally carried out using non-destructive methods. According to different detection purposes and technical principles, it can be summarized into three methods: image reflected wave and vibration.

The imaging method takes pictures or photographs of the structure's surface to see what changes have occurred in the surface morphology and determine the type of change (pattern recognition). Image recognition is the core. In the past, digital image processing technology was used, but now machine learning is used. With the development of AI, the image method is being replaced by computer vision technology. The computer is equipped with "eyes" (photograph or video camera) and "soul" (algorithms) so that the computer can perceive changes in the environment.

The reflected wave method is wave dynamics (but does not need to solve the wave equation). It is an essential non-destructive testing method applied in many industries. For example, b-ultrasonic (ultrasonic) inspection, laser measurement, sound wave detection, etc., all use the principle of wave reflection. Waves will be reflected at the junction of different mediums. The reflection distance is determined according to the wave's propagation speed and propagation time to obtain the contour (geometric form) of the interface of different mediums. Figure 6.39 shows the use of different emission waves to detect the surface geometry (such as flatness) of the road surface (or track) and the internal defects of the structure (such as potholes).

The commonly used emission waves in engineering mainly include sound, ultrasonic, and electromagnetic waves. The appropriate emission wave can be selected according to the properties of the detected object and the purpose of detection. For example, sound waves can detect defects in concrete slabs, ultrasonic waves can be used to detect defects in steel rails, and electromagnetic waves (Radar) can detect internal defects or thickness of structures. The practical use is not limited to these aspects. We need to research and expand.

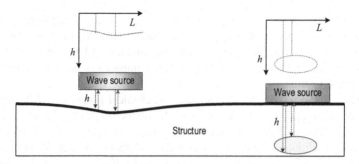

Fig. 6.39 Boundary profile determined by reflected waves

Wave is the propagation form of vibration, which exists widely in nature. The propagation of mechanical vibration constitutes mechanical waves, called sound waves in the air and called stress waves in solids. The propagation of electromagnetic field vibrations constitutes electromagnetic waves-radio waves, microwaves, infrared rays, visible light, ultraviolet rays, X-rays, etc. Any physical quantity, after being disturbed, can propagate in the form of a particular wave in the continuous medium (solid and fluid). Using the phenomenon of fluctuations, non-destructive testing can be performed. This is an essential basis for perception technology. For details, see other parts of this series.

The vibration method belongs to vibration mechanics and is a standard method for testing the mechanical properties of structures. The structure's dynamic response (displacement, velocity, and acceleration) is obtained by applying some form of power (transient shock, steady-state vibration) to the structure, and the structure's dynamic response (displacement, velocity, and acceleration) is obtained. After processing and calculation, the quantity reflects the structure's mechanical properties (such as modulus and deflection) to analyze the evolution of structure performance during the operation period. Common FWD and steady-state vibration belong to detection technology.

At present, on-site inspection technology is developed from single inspections to multiple comprehensive inspections. Various forms of comprehensive inspection vehicles are the best proof, mainly integrating the image and the reflected wave method, and the three-dimensional visualization effect has also been enhanced. In terms of comprehensive testing, there are more testing content and technologies in the railway field, and specific testing techniques may be inspiring the research and development of comprehensive testing technologies in the highway field.

4. **Intelligent inspection**

Inspection technology mainly includes sensors and data collectors. Since a micro-computer has controlled data collection, it is not difficult to realize intelligent inspection. Objectively speaking, the current intelligent detection is more like an advanced automatic detection. Still, it has already met the needs, and there is no need to force it to be linked to "intelligence."

Intelligent, comprehensive detection technology will be a good development direction, and the focus of its intelligence should be reflected in the "synthesis" of detection methods and data processing. This will be a challenging research topic and worthy of study.

> Perception technology has both commonality and industry characteristics. The so-called commonality means that the data acquisition devices are the same; the industry characteristics mean that the sensors, data processing, and analysis methods are not the same and are related to specific professional requirements. In addition, many detection methods share the same principles. For example, non-destructive testing using sound waves, ultrasonic waves, electromagnetic waves, and lasers are based on the reflected wave principle of wave dynamics or vibration mechanics. Refining the commonality from the differences helps enhance research and development.

(2) Analysis and decision-making

The main task of the maintenance phase is to make decisions on the maintenance time and scale through a comprehensive analysis of relevant information. An engineer on a computer generally does this part of the work. The "road surface management system" in the highway field is a good tool, but it needs to be developed in the direction of intelligence.

Judging from the current development of AI, the analysis and decision-making part of the work is very suitable for intelligent technology. Still, selecting different algorithms for specific situations and characteristics of different industries is necessary. Some work has achieved intelligence, but it is relatively fragmented, and a complete intelligent analysis and decision-making system have not been formed. Building such a system and running it on the platform is necessary.

1. Analysis of technology and intelligence

Information (data) analysis exists in all aspects of life, and the content involved is extensive. All signal filtering, image processing, and data statistics are information analyses. Although the specific objects faced are different, many basic principles are the same. We can learn from the successful experience of other industries, especially in intelligent analysis.

As mentioned earlier, the maintenance information is mainly composed of three parts. The three types of information do not exist in isolation but describe the performance and function of the structure from different angles. Therefore, to analyze maintenance information, it is necessary to comprehensively analyze information from different sources from a systematic perspective. Professional knowledge plays a key role, and intelligent algorithms only play an auxiliary tool.

The amount of information (modulus, material properties, gradation, etc.) transferred at the end of intelligent construction is rich, but further preprocessing is needed to form the initial state set of the structure (system) as a foundation to predict the change of state. The basic steps of intelligent analysis are given in Fig. 6.40, which can be continuously improved.

First, determine the initial state and distribution of the structure. Each state can be regarded as a homogeneous body. Second, establish an appropriate intelligent model to determine the attributes and quantities of input and output indicators. This step

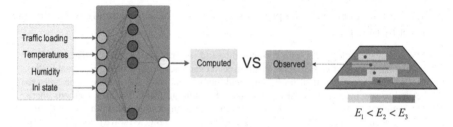

Fig. 6.40 Learning process of ANN model

is the key and needs to be enriched, and professional knowledge is needed; finally, train the model.

Figure 6.40 shows an example of training using the ANN model. The input is traffic load, temperature, humidity, and the initial state (admissible modulus) of a specific area. The output is the observable measurement (such as stress) in the area.

When the output of ANN (such as stress) differs significantly from the measured value (such as stress), the weight will be adjusted to reduce the difference until the requirement is met. Using observation data in different regions and different periods for training can expand the use of the model.

Applying the trained, intelligent model in practical work can predict the evolution of the structure state over time, as shown in Fig. 6.41.

Regarding the state of the structure, any quantity that characterizes the system can be treated as a state quantity, including modulus, density, stress, strain, etc., which can be selected as needed. If you want to get the modulus, theoretically, you should use the mechanical method for back-calculation, and it is also feasible to use intelligent prediction.

All systems must predict and analyze evolutionary behavior based on the system's initial conditions and state equations (structure) (see Sect. 2.1), and the transportation infrastructure system is the same. Under the combination of the traffic load and environment, how the state of the structural changes is the focus of the maintenance stage. However, it is not easy to establish the system state equation. It is a feasible way to seek help from an intelligent model. As for the specific operation, the model needs to be selected according to the specific situation. The above examples are for reference only. In addition, on-site inspection data can also be used as training data.

Fig. 6.41 ANN model to predict the state evolution of the structure

The above analysis of the data during the construction phase and long-term observation data. For on-site inspection, the analysis of the reflected wave and vibration methods is relatively mature, the mechanistic analysis should be the primary method, and intelligent analysis should be supplemented. Image method requires intelligent technology, such as deep learning to identify various surface defects (such as pavement cracks and track surface defects), three-dimensional visualization display, etc. This is also the current hot research direction. However, deep learning can only recognize the surface characteristics and does not know the reasons for the defects. It only plays an auxiliary role. It should be used in combination with multiple analysis methods.

In addition to the above particular analysis, it is also necessary to conduct a comprehensive analysis from a system perspective, which can be placed in the decision-making stage. All information does not exist independently, and we must be good at digging out internal connections.

For example, the unevenness of the road surface and the track is often caused by the excessive local deformation of the roadbed. If further tracked, it must be related to the filling and rolling quality during construction. The damage inside the structure is generally reflected on the surface. Mastering this analytical ability requires a solid professional foundation, which cannot be solved by intelligent methods alone. It is worth exploring whether it is possible to conduct correlation analysis by the "big data" way of thinking.

2. Intelligent maintenance decision

Analysis and decision-making are complementary: analysis is the cause, and decision-making is the result. Decision-making is based on the comprehensive analysis of maintenance information. Its main task is to decide when to repair, and the scale of repair is based on the analysis results. The early decision-making is relatively simple: to construct a comprehensive evaluation index containing multiple indexes and assign different weights to each index (determined by the scoring of domain experts). According to the size of this comprehensive index to develop a maintenance plan. As for how to construct this comprehensive index, a wealth of professional knowledge, experience, and skills are required. For example, comprehensive indicators include pavement functional indicators (smoothness), safety indicators (friction), road surface damage (cracks, ruts, potholes), and bearing capacity (deflection). Then, the above indicators will be combined in a comprehensive performance index. This is a simple decision-making method based on experience. There are many drawbacks. It is necessary to develop new intelligent decision-making methods.

The so-called decision is the decision people make to achieve a predetermined goal, and it is the third primary feature of intelligent construction. In the past, decision-making was mainly based on experience. Later, with the development of system science, operational research, computer, and behavioral science, a technical system was gradually formed, becoming one of management's cores. The essence of decision-making is optimization, and the optimal solution is selected from several alternatives according to the established goals.

Decision-making problems exist in all aspects of life, and the solutions are different, but there are certain commonalities. A decision-making Support System

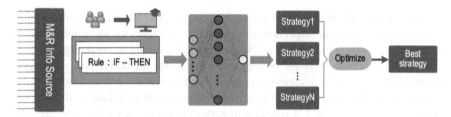

Fig. 6.42 Intelligent maintenance decision

(DSS) is a suitable means of assisting decision-making. This is a general human–machine system that uses computer technology, simulation technology, and information technology. It analyzes and judges through human–computer interaction, provides decision-makers with various reliable solutions, and supports users in making decisions. An intelligent decision-making system of maintenance can learn from the basic structure of DSS but must add the content of independent decision-making.

Intelligent decision-making is well understood: "AI + decision-making" replaces human decision-making with a machine. Therefore, the idea of realizing intelligent decision-making is elementary. First, figure out what people need to do in decision-making and see which intelligent technology can replace it.

Intelligent decision-making exists in all stages of construction with different content. In the design stage, it is the decision-making of the design plan, and in the maintenance stage, the maintenance plan is determined, which is the optimization process. You can choose the technical method of "expert system + machine learning + decision technology." The expert system uses the form of reasoning to solve some qualitative problems that cannot be described quantitatively. Machine learning solves the problem of comprehensive index prediction. Decision technology solves the problem of optimization with "cost-efficiency" as the goal, as shown in Fig. 6.42.

The decision-making in the maintenance phase is complex, and the technology involved is also more complex to be considered from the management's perspective.

6.5   Intelligent Management Technology

Management runs through the project construction process, and every link has management problems. The engineering community has become a consensus to demand quality from management and benefit from management. Technology has undergone earth-shaking changes from extensive experience management to refined scientific management, and intelligent management has been seen. This is a multidisciplinary field that deserves attention.

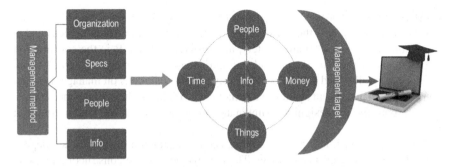

Fig. 6.43 Management methods and management objects

(1) The meaning of management

Management exists in all aspects of life, and no unified definition exists. It is roughly understood as all human organizations' planning, command, coordination, and control activity process. It effectively influences people, money, things, time, and information.

Management is a complex science that is related to human activities. The content and focus of management in different industries, departments, and levels are various. Still, there are things in common: the methods of management and the objects of management, which need to be mastered by everyone engaged in management. It is also the direction that intelligent management needs to grasp, as shown in Fig. 6.43.

There are four management methods. Among them, institutions are the main body of management, people are the core of management, and regulations and information are the basis. In the information age, except for information, management objects, such as people, money, and time, are stored in computers in the form of information (Fig. 6.43) and expressed by numbers. Therefore, modern management uses computers as tools characterized by numbers in information management. Still, the fundamental problem of management is the management of people, which is also one of the reasons for complexity.

(2) Project management

The so-called engineering management refers to a series of intervention activities (such as planning, organization, command, coordination, decision-making, and control) carried out during the entire life cycle of an engineering project (design, construction, maintenance). At each stage, the management methods and objects are different. You can refer to Fig. 6.43 for details.

For example, management is carried out in the design stage through the agency (design unit), like people, regulations, and information. Management objects are people, including designers and related personnel; finance includes design-related expenses; objects refer to design-related items; time refers to design progress; information includes various data related to design.

The management during the construction phase is the most complicated. It involves a lot of content, but the management organization is divided into the construction unit (Party A) and the construction unit (Party B). Their focuses are not the same. Party A is the construction management of the whole process and is more concerned about "quality, safety, and progress." Party B is the construction management of the bid section under its jurisdiction. It is more concerned about "efficiency, quality, and safety," and "quality and safety" focus on management and control. The management agency in the maintenance stage is the maintenance department responsible for designing maintenance programs, which has been introduced above.

The design, construction, and maintenance management information is independent of current project management. The management of the construction phase is "separate," causing uncirculated information, which is not conducive to promoting the concept of the whole life cycle. Establishing an information platform will break the phenomenon of data islands, allowing systematic coordination and overall planning.

Management is both a technique and an art. Although the methodology has commonalities, it is difficult to separate from specific industries and exist independently. Talking about management without a specific industry will be meaningless. Without a solid professional foundation, it is difficult to do an excellent job in management, which is also one of the foundations for intelligent management.

(3) **Ways to realize intelligent management**

More information will be involved if we discuss intelligence from the management's perspective. For simplicity and clarity, the essential characteristics of intelligent construction will be introduced first, followed by intelligent construction management. The current management technology has developed from informatization to networking, and intelligentization is advancing. The steps to realize intelligent management are given in Fig. 6.44.

The first is the "perception" of people, money, things, time, and information. Since project management does not exist independently, it is carried out simultaneously with project construction. Therefore, all information required for management is

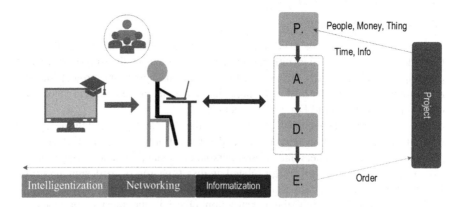

Fig. 6.44 Let machines do analysis and decision-making instead of humans

obtained through various monitoring methods in project construction and does not need to be considered separately.

The second is "analysis and decision-making," which is the focus and core of management. Management is a soft technology that analyzes and makes decisions through people (managers) in the competent authority, based on various laws and regulations and reliable information. If machines are allowed to replace humans (the purpose of AI is to replace human brain work) for analysis and decision-making, it will have the characteristics of intelligent management.

The last is "execution." In management, execution means issuing instructions based on decisions, but applying is not part of management.

Making machines analyze and make decisions is a problem solved by intelligent management. Judging from the current development, adopting the "expert system + machine learning" method should be feasible. It is necessary to clarify which tasks can be conducted using expert systems and which require machine learning, including selecting appropriate algorithms.

In the final analysis, the entire construction process is a management process. The quality and benefits of the project depend primarily on the management level. At present, intelligent management in all aspects of life is not mature, and how to combine AI and management technology well still needs in-depth research.

6.6 Virtual Construction Has Great Potential

The so-called virtual construction moves the construction process to the computer and operates on the computer, just like growing flowers and planting vegetables in video games, using virtual reality (Virtual Reality, VR) technology in engineering construction. The application will play a significant role in intelligent construction.

(1) From reality to virtual

Due to the creation of computers, people have intended to do jobs using computers, which is virtual reality. Virtual reality creates a real-world imitation (also called a twin) in a computer, upgrading computer simulation technology. Due to its excellent visual effects (three-dimensional, dynamic), this technology has attracted widespread attention from all aspects of life and has been applied in many fields. Figure 6.45 shows several application scenarios.

VR technology currently has the most applications in entertainment, education, and art, followed by military and aviation fields, and has also begun being used in construction, manufacturing, and medicine. In addition, VR technology corresponds to Augmented Reality (AR) technology, synthesizing the virtual world into the real world on the screen and interacting. It can superimpose computer-generated graphics, text, and other information into the real world.

> VR is a computer simulation system that can create and experience virtual worlds. It uses data in the real (physical) world to conduct interactive three-dimensional modeling and simulation through a computer to simulate the world in reality, called virtual reality. The

Fig. 6.45 Virtual worlds in a computer

most significant feature of VR is its direct interaction with the user. The user can directly control various parameters of the operating object, and the system can also give feedback corresponding information to the user in real-time. It also has super simulation capabilities, realizing human–computer interaction, allowing people to operate at will and get the most realistic feedback from the environment during operation. A typical application is simulation training by using VR and computer statistical simulation. The space shuttle and flight environment can be reproduced in virtual space, allowing pilots to conduct flight training and experimental operations in the virtual space. The same is true for car simulation equipment. It will significantly reduce the experimental expenses and the risk factor of the experiment. We should make good use of this technology.

The transportation infrastructure exists as an actual item in the physical world. Transforming it to (mapping) the virtual world requires a lot of data. In the current virtual technology, the primary display element is the geometric shape of the building. Its data sources are from two aspects. One is the design data from CAD; the other is the data collected by on-site laser scanning. The geometric shape determines the linear shape of roads and railways and affects driving comfort and safety. If you can simulate using the computer during the design stage, many problems can be avoided, which is the benefit of VR.

At present, this kind of technology still has significant flaws. Due to insufficient supporting data, it cannot carry the virtual simulation of the structure performance and the whole life cycle. The continuous detection in intelligent construction can obtain each structure point's performance data (modulus), which may bring a bit of hope to solving this problem.

(2) **From 2D static to 3D dynamic**

Virtual construction transforms structures from static two-dimensional graphics to dynamic three-dimensional objects, which will bring changes to engineering construction. We need to use this technology to innovate.

Three-dimensional design (Sect. 6.2) can more intuitively and accurately express geometric features. The entire design process can be carried out on a three-dimensional design platform to complete the dynamic simulation analysis of the relationship between geometry and traffic, drainage facilities, environment, landscape, etc. This can be used to timely find defects and inconsistencies in the design and reduce design changes, which cannot be completed in two-dimensional designs.

After virtual construction changes, the construction from static to dynamic display dramatically increases the visualization effect. This can allow dynamic simulations in the design stage and visual construction in the construction stage. Using a three-dimensional digital model, the geometric shape of the structure can be viewed in the VR environment to help the construction staff better understand the situation. For buildings, you can also view the internal and external structures. Models can also be viewed using head-mounted equipment, and project participants can collaborate remotely. Different majors can be immersed in the same environment to collaborate, communicate, and open up communication channels between the virtual and physical worlds.

(3) Emphasis and development of ideas of VR for filling projects

Virtual reality integrates many technologies such as three-dimension dynamics, simulation, vision, and sensors. Many industries desire to carry out various tasks in a virtual environment. The external shape's virtual and dynamic simulation analysis of road and railway engineering is essential. Still, the simulation of the performance and function of the structure should also be indispensable. It is also an indispensable focus of the entire life cycle but is more complicated. Road engineering is taken as an example below for a brief description.

Due to different engineering qualities, various damages will occur during the entire life cycle of roads, from construction to scrapping. The repairs will vary depending on structural forms, material selection, and construction techniques. Many problems will be avoided if these results can be predicted during the design phase. How can we realize this wish? Using technologies, such as virtual reality, to simulate a road on the computer and perform analyses of the virtual road will be a solution. In this regard, we can learn from the successful experience of other industries.

Aircraft design has moved from two-dimensional to three-dimensional, from physical modeling to virtual prototypes in the aviation field, combining with various industrial and data management software to realize informatization, networking, virtualization, and intelligence. Engineers can perform part design and product assembly on the computer, complete various analyses and calculations, perform assembly inspections and motion simulations, etc. The production process of virtual products is almost the same as the actual process. Three-dimensional models describe all parts to meet the needs of design, manufacturing process, and life cycle management, which significantly improves design efficiency and economic benefits. The virtual construction of road engineering can also be developed through this approach. Still, the road is a filled structure, which is quite different from the assembled structure of an aircraft and cannot be copied thoroughly and needs another way.

The performance and function of components of all assembled structures, including airplanes, can be determined in advance. Therefore assembly testing and analysis can be carried out on the virtual prototype. However, the road is a filled structure; the "components" are subgrade, base, and surface. The completed structure cannot be guaranteed to match the design requirements. Its performance and functions are difficult to determine in advance, and the simulation is useless. It is necessary to use the basic knowledge of intelligent engineering design (Sect. 6.2). This would take advantage of massive and historical data, establish the association between materials, structure, craftsmanship, and quality, form an initial component library, and accumulate intelligent construction by adding the new data to improve the library. This way, the information in the component library can be used to assemble, test, and analyze the structure. Figure 6.45 shows some of the main content of virtual simulation of road engineering.

In Fig. 6.46, the external geometry contains much information. Geometry simulation is the most important one related to driving comfort and safety and corresponds to the landscape. It is more suitable for immersive simulation of wearing equipment to feel the driving and environment to improve the design. Other simulation aspects include traffic engineering, drainage, and slope protection.

The construction process simulation (virtual construction) is different from the assembly simulation (virtual manufacturing), which still needs to be studied, and the roadbed simulation is the most difficult. Because the "components" of the structure are all built on-site, they cannot be manufactured in advance, nor can they be determined for their performance after molding. There is still variance in material preparation, transportation, paving, and rolling. The virtual simulation of this part still needs a lot of empirical data as support.

For example, in the construction simulation of graded crushed aggregate, it is necessary to set the gradation and particle shape and generate a filling layer with a certain thickness. Then, according to the corresponding relationship between rollers-process-quality, different rollers, and different rolling process parameters, various combination simulations are performed for the corresponding rolling quality. Regarding the correspondence relationship between rollers-process quality, a large amount of data must be collected (See Sect. 6.2).

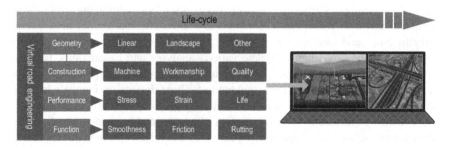

Fig. 6.46 The main content of the virtual road engineering

If the information in the component library is incomplete, it is of little practical significance to carry out construction simulation in the design stage. If the intelligent construction method is adopted, the quality information of each point of the filling layer can be obtained, and the simulation in the maintenance phase is of practical significance. It can be seen as the foundation for preventive maintenance.

Performance simulation is conducted for the structure. Suppose the physical parameters (modulus, density, Poisson's ratio) and distribution of the structure are known. In that case, each structure's stress, strain, deformation, and fatigue life under the driving load can be calculated. It mainly involves the mechanic theory of interaction between the vehicle and road surface, which is very mature.

Functional simulation is conducted for the road surface, also called virtual simulation of the surface function, which involves the smoothness, anti-skid ability, and rutting of the road surface. The simulation's purpose is to analyze these diseases' impact on driving comfort and safety. The data of pavement surface function collected from on-site inspection during the maintenance phase involves the interaction between the vehicle and road surface, which is considered from the perspective of comfort and safety and involves knowledge of the vehicle's random vibration and dynamics. In addition, the effects of water and temperature on surface functions should also be considered to make the simulation more realistic.

At present, virtual construction is still in the early stage of development. It is usually seen as the three-dimensional visualization effect, but the essence is different from the perspective of engineers. Virtual construction is part of the cyber-physical system (CPS, Sect. 5.7), composed of digital models, status information, and control information. The design and construction process of the structure can be verified in the virtual world in advance. On the other hand, according to the structure's state of the construction and maintenance process, it can be reproduced dynamically in real-time in the virtual environment for easy analysis and control.

With advanced technology development, many industries are inclined to use high-performance simulation to replace expensive physical experiments, saving costs and shortening the development cycle. At present, virtual simulation has become an indispensable means of product design, and the level of simulation has gradually developed from the macro to the micro. Intelligent construction should pay close attention to the development trends of related industries and technologies. The gap between different disciplines is narrowing, and multidisciplinary integration is emerging.

6.7 Intelligent Construction Example: Intelligent Compaction

The previous analysis of the intelligentization problems in each stage of construction explains the overall concept of the development of intelligent construction. Judging from the current situation, the integration of many intelligent technologies and engineering construction has just begun. Intelligent compaction, the core technology in intelligent construction, is the fastest developing and relatively most mature. A preliminary introduction has been made in Sect. 6.3. Here, intelligent compaction is an example of intelligent construction to show the process of "perception, analysis, decision-making, and execution," which helps understand the connotation of intelligent construction. The detailed discussion of intelligent compaction is in other parts of the series.

(1) **What is intelligent compaction**

Intelligent compaction comes from continuous compaction control, a technology that uses the response information of the roller during the compaction process to control the compaction quality continuously.

The so-called intelligent compaction refers to the continuous acquisition of compaction quality control information according to the perceived response signal of the vibrating wheel of the roller during the rolling process. Through independent learning of control information, filler information, rolling process information, independent analysis, decision-making, and feedback control of compaction quality are applied to improve compaction quality. Figure 6.47 shows the basic principle of intelligent compaction.

Intelligent compaction is a comprehensive application of dynamics (vibration mechanics and wave dynamics). According to different technical principles, intelligent compaction is divided into five levels, and advanced technology uses levels three and above mechanical control indicators. Advanced intelligent compaction should meet four basic requirements. First, the roller must be equipped with a "brain" or control system, including a measurement system (vibration sensor + data collector), control software, algorithm software, etc. Second, the most critical element is a control index reflecting the compaction quality to obtain reliable control information. Third, the control system must be able to learn, reason, analyze, make decisions, and issue feedback control instructions; fourth, the roller must automatically adjust process parameters to realize intelligence in performance and function.

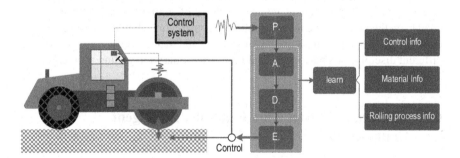

Fig. 6.47 Principle of intelligent compaction

Some advanced control systems have the essential characteristics of "perception, analysis, decision-making, and control." Some rollers also have automatic feedback control. Their combination will lay the foundation for the further popularization of this technology.

(2) The work process of intelligent compaction

The working process of intelligent compaction is not complicated, which also confirms that the more intelligent object is more straightforward to operate, which can be mastered according to the black box method. But the technology of intelligent compaction is not simple and involves more disciplines. To engage in research and development, you should master this knowledge.

1. Perceive compaction quality information

The first step of intelligent compaction is to perceive compaction quality information, which is a prerequisite. Due to the mobility of the roller during the compaction, the current control system can only continuously measure the vibration response signal of the wheel. It is necessary to analyze the information related to the compaction quality (control index) based on the response signals. Fortunately, the material properties of the compacted body (modulus and density) are obtained, which is crucial for intelligent compaction technology.

As mentioned in Sect. 6.3, the rolling process involves the roller and compacted body. The essence is the interaction between the rigid cylinder and the compacted elastic material, which a classical mechanic problem can describe. Figure 6.48 shows the technical principle and example equations of intelligent compaction on elastic material.

Intelligent compaction needs to identify the modulus and density (or other indexes) of the compacted body according to the response signals of the vibrating wheel. The

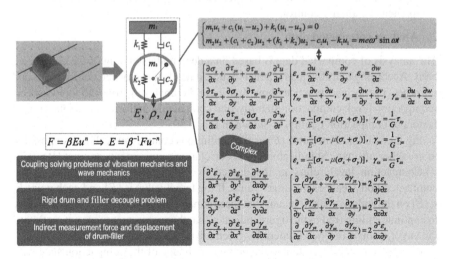

Fig. 6.48 Dynamic principles and technical difficulties of intelligent compaction

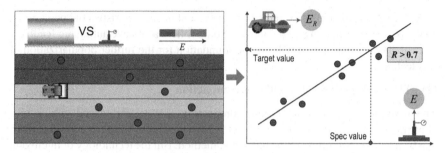

Fig. 6.49 Comparison tests were used to verify the accuracy of the control index

dynamic model shown in Fig. 6.48 has no precise analytical solution. The existing solutions are approximate, but they can satisfy the accuracy requirements. When the steel wheel bounces, the impact model must handle the non-coupling interaction with the compact body. Due to the roller movement, only the responses of the vibrating wheel (acceleration, velocity, and displacement signals) are directly measured during the rolling process. Other methods need to be used to obtain the information required for calculation.

The accuracy of the control index obtained according to the above model needs to be verified, and the comparison test is generally used for verification. Figure 6.49 shows comparative tests and data processing methods and requirements.

The correlation coefficient (R) determines the consistency of the intelligent compaction control index and the point detection index (modulus or density). The R-value must be greater than 0.7. In addition, according to the qualification of point detection, the target value of intelligent compaction control can be obtained.

The obtained data will cover the entire compaction area if the control system can continuously perceive the compaction quality information, laying the foundation for comprehensive quality control.

2. **Analyze compacted data**

The analysis of compacted data is based on a deep understanding of the compacted body. According to traffic load characteristics, it needs to be carried out for roads and railways. Theory and practice show that it can be analyzed from the three aspects (compaction level, compaction stability, and compaction uniformity) and the control factors of compaction quality, as shown in Fig. 6.50.

Figure 6.50 takes modulus (E) as the continuous control index. It is also possible to use other mechanical indicators if the correlation requirements are met ($R > 0.7$).

Compaction level refers to the degree to which the measured value of the compacted body reaches the target value. If the target value is $[E]$ and each point E_i < $[E]$ on the rolling track, it means that the compaction is not over yet. It is necessary to continue rolling (red in the figure indicates that the requirements are not met) until the target value is reached (generally, green means qualified).

Compaction stability refers to the degree of change of the compaction state with the number of rolling passes and is generally expressed by the relative error of the

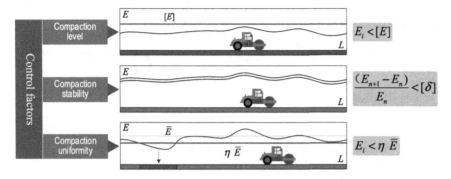

Fig. 6.50 Control elements and analysis methods

measured values on the front and back of the same rolling wheel track. The current difference between the last two passes is minimal, indicating that the compaction of the roller under the process parameters has been fully utilized, and continued rolling will cause "over-compaction."

Compaction uniformity refers to the uniformity of the distribution of the performance parameters of the compacted body on the rolling surface, and the corresponding is the variability (including the segregation of materials). The uneven compaction is one of the reasons for the early damage of the superstructure and must be controlled. There are many ways to characterize the uniformity of compaction, and it is used to decide whether it is uniform. When $E_i < \eta \cdot \bar{E}$, it means uniformity. Where, $\bar{E}$ is the average value of the measured values; η is the coefficient of unevenness, and $\eta = 0.80$ for a high-speed railway in China.

3. **Decision-making on the compaction process**

Intelligent compaction is mainly based on the analysis results to decide whether to continue compaction, compact (adjust the compaction process) and issue control instructions.

If the packing is good, the analysis and decision-making of the compaction data are easy to carry out without too much intelligent technology. As long as Fig. 6.49 uses the method, ordinary control systems can automatically accomplish this task.

Suppose there is a problem with the packing or compaction process (the combination of vibration quality, vibration frequency, and excitation force of the roller). In that case, it is necessary to use intelligent technology for analysis and decision-making, including the analysis and decision-making of the compaction process, and the packing can be compacted. If the control system masters these skills, it is necessary to select appropriate intelligent algorithms and use known data to carry out a large amount of training (learning) on the control system of the roller. Expert systems can also be used, and they can be combined for extensive use.

4. **Feedback control**

The purpose of analysis and decision-making is to control the quality of compaction. The control factors are the degree of compaction, compaction stability, and

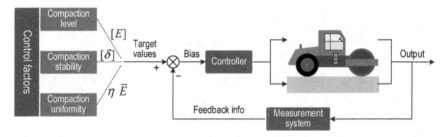

Fig. 6.51 Principle of compaction quality feedback control

compaction uniformity; the controlled objects are the filler, the roller type, and the process parameters. Figure 6.51 shows the primary compaction quality feedback control (see other series parts). A little more explanation is as follows.

The basic idea of feedback control is to compare the feedback information with the set value (target) and adjust the relevant parameters according to the magnitude of the deviation to minimize the deviation. The set values of compaction level, compaction stability, and compaction uniformity are shown in Fig. 6.51. The measurement system gives the corresponding feedback information (part of the control system).

The mechanism that adjusts the parameters is called a controller. This mainly refers to the mechanism that adjusts the process parameters (speed, vibration frequency, excitation force, etc.). It can also be the technical means to adjust the properties of the filler (particle shape, gradation, water content, etc.), which is called a "soft controller."

If it is an intelligent roller, the above control behavior will be completed automatically without human intervention. If it is an ordinary roller, the control system will issue instructions for execution through the display, manually adjust the roller's mechanism or improve the packing's properties.

The above is the process of intelligent compaction, which fully embodies the four primary characteristics of intelligent construction. Nevertheless, many research gaps exist in models, hardware, software, and algorithms.

(3) **Visualization of compaction results**

The data obtained by intelligent compaction is vast, and it is an excellent choice to display it visually. The current control system can perform real-time and online visual displays during the rolling process and remote displays through an information platform. All compaction results are observed easily for convenient quality control and management. Figure 6.52 shows typical visualization results: compaction state distribution and compaction degree distribution. Both were generated using continuous control data (modulus).

In Fig. 6.52, the target value of the level of compaction control is $[E] = 80$ Mpa, by which the map is divided into two areas, red and green. The distribution of compaction state shows the distribution of the modulus of each point at the compaction surface, distinguished by different colors from low (red) to high (green), and has more information than the distribution of compaction degree.

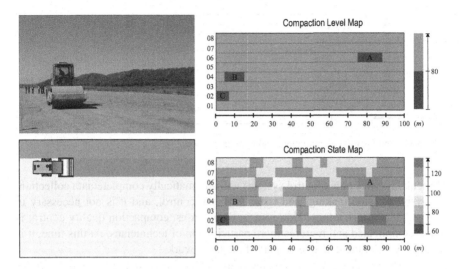

Fig. 6.52 Visualized display of compaction control results

For example, in the distribution of compaction degree, the red areas are A, B, and C, indicating that it is less than 80 MPa, but the specific value is unknown. But in the distribution of compaction state, the specific values of the A, B, and C areas can be known. The critical point is that traditional spot detection can be performed in low-value areas, avoiding random sampling.

(4) **Coordination of continuous and point inspection**

Intelligent compaction can be regarded as a continuous detection with the roller as the loading device. Due to the different models and parameters of the roller, the loading method and weight are inconsistent, and some information cannot be directly measured. The detection accuracy is not too high. This accuracy can still meet the process control or inspection technology requirements. Point detection can be used for high-precision detection, but the random sampling method must be changed to fixed-point detection, which requires continuous detection results.

Figure 6.53 shows the combination of continuous detection and point-level detection. Continuous testing can determine the distribution of compaction state at the paved surface (the red area represents the low-value state, and it has nothing to do with whether it is qualified) and select several low-value areas for point level detection. If the point inspection is qualified, the entire rolling area is qualified. If the point level inspection is unqualified, the low-value area can be sorted and tested until the qualified area is found. Then, the total area is processed.

(5) **Comprehensive use of intelligent technology**

In the intelligent compaction technology introduced earlier, "intelligence" has not been applied extensively. As long as the filling properties are good and the engineering machinery is suitable, compaction quality control using the method in Fig. 6.49 is not

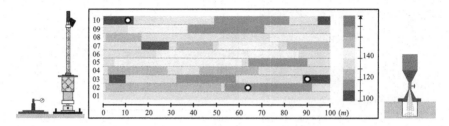

Fig. 6.53 Combination of continuous and point inspection

difficult. The machine (control system) will automatically complete data collection, analysis, decision-making, and issue feedback control, and it is not necessary to use intelligent technology. However, in many cases, compaction quality control is more complicated and requires more participation of technicians. At this time, it is necessary to let the machine do part of the labor work.

The rolling process is a complex non-linear interaction between the roller and compaction. Compaction quality is affected by material properties and the compaction process. This effect is also reflected in the response signals of the roller and eventually refected by control indicators. Due to the filler properties' complexity and the roller performance diversity, the rolling work is not easy in many cases, and many unexpected situations may occur.

The practice has shown that, based on the change of the control indicators and some necessary information, the compaction quality can be analyzed, the properties of the filler and the compaction process can be analyzed, and control methods can be proposed. It is intelligent to give this kind of work to the machine, but it needs to learn and requires a large amount of known data for training. The following describes the application of several intelligent technologies in compaction control.

1. **Artificial neural network (ANN)**

ANN is a supervised learning algorithm that requires much-known data for training. A simple application of ANN is to establish the relationship between intelligent compaction and point level detection by comparing test data, which is better than linear regression.

Another application of ANN is to judge whether the compaction quality is qualified, which is a classification problem. Figure 6.54 shows the basic principle of the algorithm. Of course, SVM can also be used for processing and providing more accurate predictions.

The input of ANN consists of three components: the performance data (such as modulus) obtained by continuous testing, the roller parameters (including the quality of the roller, the vibration quality, the vibration frequency, and the exciting force), and filler properties (including the type and gradation of the filler, particle shape, and water content). There are two outputs: "1" means qualified, and "0" means unqualified.

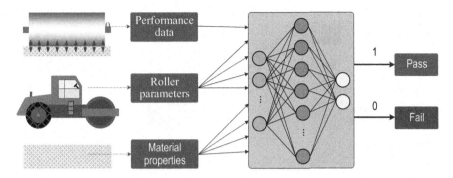

Fig. 6.54 ANN for classification of compaction quality

A multi-classification algorithm to divide the compaction quality into several categories may be better. The binary category classification is just an example to explain the idea.

The above ANN model needs to learn a lot of known data from road tests and historical records. The data collected by intelligent compaction will be the primary source in the future.

2. **Expert system and hybrid expert system**

The expert system has been mentioned many times before. Although it has no universal applicability and learning ability, it is still meaningful in the professional field. Once integrated with machine learning, it will have a variety of abilities. It is a promising intelligent technology in the field of construction.

Compaction quality control involves packing, rollers, and control standards. Some of them can be described quantitatively, such as the standards of compaction degree, compaction stability, compaction uniformity, and compaction process parameters. Some can only be described qualitatively, such as properties of the filler (gradation and segregation). These can be transformed into rules (knowledge) in the expert system's "IF–THEN." Please refer to the example given in Fig. 3.1.

For example, the rules for describing the compactability of coarse-grained materials are: IF coarse-grained materials are not well-graded and THEN are not easy to compact. The rules for describing the qualified degree of compaction are: IF [E] = 80 Mpa, AND E = 100 Mpa [E], THEN is qualified.

If there is enough knowledge in the database of compaction, the expert system can be applied to deal with various rolling problems based on the description of the specific problem. After reasoning, the system will conclude (answer) on how to solve the problem.

The expert system's knowledge is obtained from domain experts, but its shortcomings are also prominent (see Sect. 3.1). To solve expert systems' shortcomings, combining machine learning with expert systems is feasible, known as developing hybrid expert systems. For compaction control, combining multiple technologies may lead to better development of intelligent compaction technology. Figure 6.55 shows the schematic diagram of the neuro-expert system.

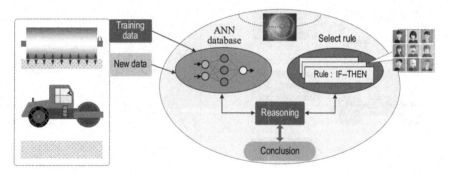

Fig. 6.55 Neuro-expert system

In the neuro-expert system, part of the knowledge is explained by the trained ANN, and the knowledge is stored in the weight of the neuronal connection, an implicit expression. An ANN model can explain relevant knowledge (a group of rules), automatically obtain a new weight distribution, and update the database according to the new training data. For example, the ANN model in Fig. 6.55 can be viewed as a set of rules.

IT-THEN can still be used for rules that are difficult to represent in ANN. A fuzzy-expert system can be used for fuzzy problems, and a probabilistic expert system can be used for non-deterministic problems. In addition, it is also worth studying whether the deep learning emerging in recent years can be combined with the expert system.

3. **Reinforcement learning**

The intelligent technologies described above require a large amount of training data and involve many field tests and the experience of experts at a very high cost. In the absence of training data, other algorithms may be considered. Reinforcement learning is a class of dynamic decision and control algorithms suitable for compaction control problems.

According to Sect. 3.5, the subject is a road roller in reinforcement learning, and the environment is a filling body. The effect of the subject on the environment is rolling, which can change the current compaction state of the filling body, and the change of this state can be observed by the roller giving the roller a profit (reward). As a reaction to the rolling quality, the bigger the profits better. According to the two feedback information, reinforcement learning guides the roller to make the most favorable rolling action to obtain the maximum benefits, as shown in Fig. 6.56.

Reinforcement learning emphasizes "learning by doing, doing while learning." The score is deducted (negative income) if the decision is wrong, and the reward is added when correct. The roller can learn to choose what kind of craft (best strategy) under what circumstances by constantly adjusting the action (compaction technique) justified by trial and error to get the best reward. Since it is difficult to express the packing in a fixed model during the rolling process, a model-free algorithm (Sect. 3.5) can be considered, and continuous exploration (trial and error) can be used.

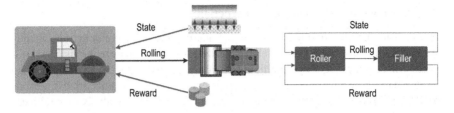

Fig. 6.56 The idea of using reinforcement learning for compaction control

Reinforcement learning is a Markov process, and this learning method has some problems. Another algorithm worth paying attention to is Generative Adversarial Network (GAN). This algorithm uses the adversarial training mechanism to train two neural networks (generator and discriminator), avoiding problems caused by repeated use of the Markov learning mechanism. Whether GAN can be applied is worth studying.

The intelligent compaction technology was briefly introduced, mainly focusing on rock and soil compaction control. The intelligent compaction of asphalt mixture involves the coupling effect between the two steel wheels of the roller and the influence of temperature. The situation is more complicated, and it is still under development.

6.8 The Risks of Intelligent Construction

Intelligent construction is a new thing and involves "intelligent technology." Given the various risks and ethical issues in the AI field, here is a brief discussion of security risks related to intelligent construction.

(1) Risks in engineering construction are everywhere

Risk is ubiquitous and always observed in construction, and intelligent technology will not cause new risks. For example, there are various risks to personal safety and operation errors in surveys and measurements. There are risks of calculation errors and drawing errors in design, and there are more risks in construction. Faced with these risks, have we ever worried too much? The answer is no!

Construction risks can be seen everywhere, but no one has ever worried too much. These can be avoided by relying on various regulations, systems, and advanced technologies (including intelligent technologies). But why do people start to worry about various risks regarding intelligence-related issues? The reason may be related to the excessive rendering by the media and science fiction movies, causing humans to mistakenly believe that robots want to rule the world and humans will become slaves. If you understand the nature of machine learning algorithms, will you still think so?

(2)   **Possible risks in intelligent construction**

As a new thing, there are some risks in implementing intelligent construction, but they are not like the ones described in science fiction movies. To analyze the risks, we still need to discuss the essence of intelligent construction: using machines (computers) to replace part of the human brain's work. Since computers can be embedded in various objects, the risk is brought by using the computer. The risk analysis of intelligent construction should be carried out. The data risk (security) issue has been described in Sect. 4.7.

Assuming that computers will bring risks, the first thing that comes to mind must be security risks, such as viruses, hackers, etc., which may interfere with computers to steal information, remote control, or even destroy. The computer risk is the software risk. One is deliberately done by programming. This is the case of virus software and hacking software, but it is a crime and is not within the scope of discussion here. The other is the "BUG" in the program code, the most common. It's not easy to altogether avoid the risks.

There are certain risks in any scenario where computers replace human work for intelligent construction. In addition to the above risks, the program "crash" is also a common problem caused by bugs. How to prevent and control them is an issue for programmers. In addition, the following issues are worth discussing in popularizing intelligent construction technology.

1.   **Can the machine make the decisions?**

Intelligent construction can allow machines to replace humans to analyze the perceived data and make decisions. The real question is, do you dare to use these machines to analyze and make decisions?

With the current technology, the suggestions made by machines will undoubtedly be reviewed by humans. For a long time, machines will still assist humans in their work.

2.   **Is unmanned construction feasible?**

Construction robots and automated machinery are becoming a hot topic. According to the current technical level, it is easy to apply unmanned driving. The key is that the construction machinery must have a sound perception system and positioning system to replace the driver's vision (observation) and hearing (speaking) and accurately determine the working position. The construction site (worksite) is different from the road in operation. The use of construction robots will not cause too many safety problems. This is different from the unmanned driving of cars, but various rules and regulations should be formulated, and related technologies should be improved and standardized.

In some dangerous areas, like steep slopes, construction robots (such as unmanned excavators) can work, which will be safer. If it is not combined with intelligent compaction technology for paving and rolling operations, it is of little practical meaning and will only become a new concept. Because the roller needs to know the quality of rolling in real-time and learn to change the parameters of the compaction process, only relying on controlling the number of rolling passes is insufficient.

3. What should I do if the network or power is cut off?

With the development of technology and the improvement of automation and intelligence, human beings will rely more and more on machines. However, what problems will be caused by excessive reliance is worth discussing. What should we do if the world is out of power and the internet? Due to some malfunction, the construction robot's sensing and positioning systems have no signals during the construction process. Can the machine still work correctly? Many similar issues are no longer within the scope of intelligent construction and require collaborative work with other domains.

Chapter 7
The Road to the Future

Abstract This chapter first analyzes the main goals of current intelligent construction. It then analyzes the key to realizing road and railway informatization, forecasts the future development prospects of transportation infrastructure, and discusses the application of intelligent technology, including intelligent construction in various fields. The basic features and implementation steps are summarized, the relationship between related technologies of intelligent construction is sorted out, and the relevant theoretical basic knowledge is given.

7.1 Quality is the Key

There are two types of "products" for intelligent construction: one is intelligent products, and other is high-quality products. Access to intelligent products (roads, railways, etc.) is undoubtedly good, but it's not quite there yet. The main goal at this stage is to improve the quality of the project and get high-quality "products" (broadly speaking, including all infrastructure). The meaning of engineering quality is multi-faceted. The internal quality includes the project's function, durability, reliability, applicability, etc. The external quality includes the beauty of architectural art, engineering technology, the coordination of ecology, the integration of culture, and post-service. In our case, intelligent construction mainly focuses on improving internal quality.

From the perspective of engineering construction, quality is mainly reflected in engineering quality. In the whole life cycle, the minimum number of maintenance and the minimum maintenance scale (Fig. 7.1) are achieved, and the performance and function are satisfied. They are driving safety and comfort needs. Only on this basis can we pursue intelligent transportation infrastructure.

To improve project quality, we must start from the source. Since the "trapezoid" shape of roads and railways will not change, and the structural form is challenging to change, such a source is the raw material, which is the first hurdle to grasping the quality. The second level is construction, that is, the manufacturing process of "products", which is the key to good quality and the main battlefield of intelligent

© China Railway Publishing House Co., Ltd. 2023
G. Xu and D. Wang, *Introduction to Intelligent Construction Technology of Transportation Infrastructure*, Springer Tracts in Civil Engineering, https://doi.org/10.1007/978-3-031-13433-3_7

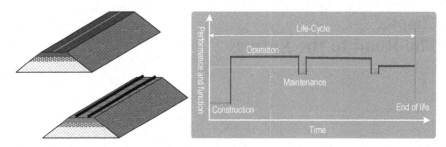

Fig. 7.1 Improved quality results in reduced maintenance

construction. At present, intelligent construction is mainly carried out around this part, and it is also the focus of attention of all parties.

With the popularization of intelligent construction, how to further improve the project quality and build a high-quality road without increasing the cost is an examination question for intelligent construction. Let us wait and see.

7.2   Informatization of Roads Depends on the Cost

"Informatization of roads" mainly refers to roads and railways (including airports and urban roads) that can provide information on the performance and function of structures to the outside world and is a general term for information roads and information railways. The concept of information highway has long existed, such as Britain's Traffic Information Highway (TIH), Video Information Highway (VIH).

At the same time as intelligent construction, various sensors (Sect. 6.3) can be installed on infrastructures (items) such as roads and railways so that they have the capabilities of perception, touch, and vision and can perceive the internal performance information and external environment information of the structure. They can transmit information to the outside world (information platform), which is the concept of information highways and information railways, mainly used to monitor changes in the performance of structures. Currently, "Smart+" is mainly used to call it (not wisdom). Strictly speaking, the current information road has not yet reached the level of having too much intelligence, and it is even farther away from intelligence. Even so, there is no need to tangle too much on terminology. The key is to look at the substance—whether it can provide various information to the outside world.

At present, the "informationized" highways and railways mainly have the ability of perception and data communication. They can perceive the relevant information of the internal and external environment of the structure and transmit it to the outside world. Still, they cannot analyze and make decisions. Autonomous learning belongs to the perceptual layer (Sect. 5.7), and feedback control and self-repair of the structure are out of the question (that is the way of the future). Learning, reasoning, analysis, and decision-making are all done on the information platform (belonging to the

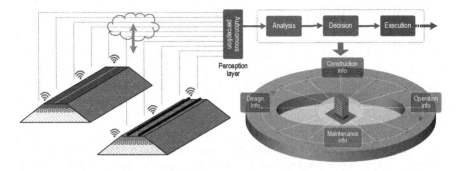

Fig. 7.2 Technical characteristics of information highway and railway

application layer) and have little to do with the perception layer, as shown in Fig. 7.2. If it is a CPS platform, it can also realize remote control. However, it still needs to be studied about the remote control targets.

As stationary objects, road and railway structures can achieve intelligence as long as microcomputers and intelligent algorithms are embedded in the structures. However, due to the limited processing capability of the single-chip microcomputer, the level of intelligence is limited, far less than the processing capability on the information platform.

From the perspective of IoT, information highways and railways belong to the perception layer. The main task is to perceive how the internal performance of the structure changes under the combined action of the traffic load and the natural environment. It does not require on-site analysis and decision-making. The data is comprehensively processed and analyzed on the information platform to provide a basis for maintenance decisions. Various analyses can also be performed in a virtual environment (Sect. 6.6) as long as the information is sufficient. Therefore, the practical significance of whether the perception layer has intelligence is not great, mainly because the perception terminal needs to automatically provide the information platform with the performance data of the structure.

Roads and railways provide vehicle services from the functional aspect of transportation infrastructure. Since the train runs along a fixed track and the operation is relatively simple, the current research focuses on intelligent transportation (ITS) on the highway, which is the case with full-fledged autonomous driving. Arranging various sensing elements on the road surface and surrounding areas can assist the car in driving, give early warnings to driving conditions or road conditions, and improve traffic capacity.

It is very easy to "informatize" ordinary roads and railways. As long as various sensing terminals are arranged inside, on the surface, and around the structure, and wireless communication capabilities are available, it can be realized, but the cost is very large, and the key is to look at the investment. The amount of funds and the cost will control the number and technical level of sensing terminals.

7.3   The Path Forward is Innovation

"The road of the future" refers to the road of intelligence, the road of wisdom. The so-called intelligent road refers to roads and railways (structures) with autonomous "perception, analysis, decision-making, and execution" capabilities. "Execution" here mainly refers to the ability to self-heal and control. The road of wisdom is farther and must be reflected in creativity.

Self-diagnosis is easy to understand. That is, the structure can autonomously perceive changes in performance and status. After autonomous analysis and decision-making on data, it can conclude whether it needs repair or control and issue corresponding instructions. Self-diagnosis is the integration of "perception, analysis, and decision-making" in the essential characteristics of intelligent construction, as shown in Fig. 7.3.

With the development of science and technology, making the structure have autonomous perception, analysis, and decision-making capabilities is not difficult. As long as a new type of sensor (such as MEMS) integrates data acquisition, programming, and data processing capabilities embedded in the structure, an appropriate intelligent algorithm can be implemented. The difficulty lies in making the structure self-heal and self-adaptation, which current materials and structures cannot achieve. Therefore, it is necessary to expand research ideas and technological innovation.

We know that living organisms, including humans, can generally heal themselves after injury. Inspired by this, the emerging smart material is a material with self-perception, self-diagnosis, and self-healing capabilities, which is the fourth generation of materials after natural materials, synthetic polymer materials, and artificially designed materials. Understanding the characteristics of smart materials will help us develop new materials and structures suitable for roads and railways.

Intelligent materials are three kinds of raw materials with sensing, action, and processing functions embedded in the material matrix. Polymer materials and lightweight non-ferrous alloys can be intelligent materials. "Sensing" materials can detect external changes such as stress and temperature, such as piezoelectric materials, optical fiber materials, etc. "Action" materials can generate more strain and

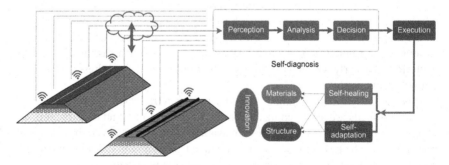

Fig. 7.3 Characteristics of the intelligent road and intelligent railway

stress, suitable for responses and controls, such as piezoelectric materials, electrorheological materials, etc. Finally, "processing" materials mainly include conductive, optical fibers, and semiconductor materials.

Intelligent materials also have the functions of feedback, recognition, response, self-diagnosis, self-repair/healing, and self-adaptation. Self-repair/healing repairs damage through regeneration mechanisms such as reproduction, growth, and in-situ recombination. Self-adaptation refers to materials that can change according to the external environment by automatically adjusting their structure and function and changing their state and behavior to respond appropriately to external world changes.

The intelligent structure is a bionic structural system that integrates the main structure, sensors, controller, and driver. It has self-diagnosis, self-monitoring, environmental self-adaptation, self-repair/healing characteristics, and intelligent functions. It is generally compatible with intelligent materials. The structural forms of highways and railways subbase are layered systems formed by paving and compaction. It is worth thinking about whether new structural forms may adapt to structural deformation in the future.

Intelligent materials and intelligent structures have been used in civil engineering. For example, self-healing fibers added to concrete can sense cracks in concrete and corrosion of steel bars and automatically bond concrete cracks or prevent corrosion of steel bars. Intelligent materials in buildings can sense real-time changes in the external force, temperature, and cracks on the material itself and identify the damage to the material. Intelligent materials can also use their adaptability to warn hazardous situations early and reduce/eliminate hazards through prediction, adaptive adjustment, and self-repair. The above examples may inspire intelligent highways and railways research and development.

There are similar intelligent materials/structure research for highways. Examples include self-repair/healing of asphalt materials, self-melting of snow, self-adjustment of subgrade humidity, etc. Other examples include compression-resistant solar panels on road surfaces for electricity generation and compression-resistant LEDs for lighting. However, there is still a long way to go for field applications. Innovative materials, technologies, and theories from multiple disciplines will help realize the goals of intelligent highways and railways.

The road of the future is the road of intelligence and wisdom. Many new materials, new technologies, and new theories will be of great use, and the integration of multiple disciplines will be further strengthened. It's a promising territory and a testing ground for innovation that deserves a good grasp.

7.4 The Future of ICT is at Our Doorstep

Like AI, the rise of intelligent construction and "AI+" is becoming an increasingly hot field. At this time, we should calm down and think about it. What is the purpose of intelligent construction? Are those popular technologies at its core? Can AI really improve engineering quality? Although there are many doubts, dare to face these doubts.

Most of these doubts are related to our high hopes for AI, and have a lot to do with one-sided media reports and various commercial hype. Practically speaking, there are very few "AI+" projects that have actually landed at present. The so-called greater the hope, the greater the disappointment. Intelligent construction cannot be built on a castle in the air, but it is still necessary to do things down-to-earth. The following briefly summarizes the technologies and theories involved in intelligent construction. As long as you understand these, you will have the desired answers to the previous doubts.

Intelligent construction integrates a variety of technologies (AI, electronics and information, dynamics, data science, automatic control, cloud computing, Internet of Things and professional technology, etc.), which looks dazzling but is not chaotic at all. The context of interdisciplinary integration is still clear. As long as it is a simple combing process, you can see more clearly. Doing so is not only conducive to mastering the relevant technologies of intelligent construction and engineering applications, but also conducive to the development of various technologies.

We know that the main purpose of developing AI is to replace the human brain with computers for mental work, but the application of intelligent technology (such as intelligent construction) is by no means a simple combination of "AI + professional technology", but requires the integration of multiple technologies. In order to straighten out the various technologies involved in intelligent construction, it is necessary to start from the basic characteristics of intelligent construction, which is also the implementation steps of intelligent construction (including other "AI+"), as shown in Fig. 7.4.

Regardless of the discipline, the basic steps for people to do things are the same, but the specific content is different. The same is true for machines. Therefore, the basic steps of replacing humans with machines can be summed up as "perceive, analyze, decide, execute—P.A.D.E." (Fig. 7.4). These four steps are proposed on the basis of summarizing the projects done by human beings. They are universal to a certain extent and are the basic route for the application of intelligent technology in various

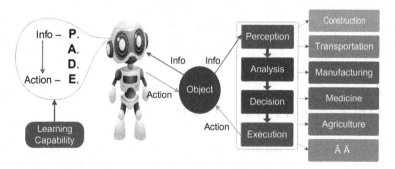

Fig. 7.4 Characteristics and implementation procedure for "AI+"

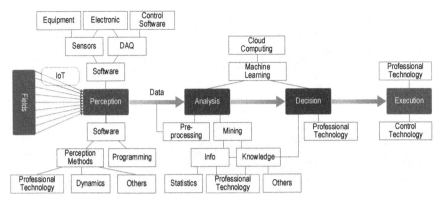

Fig. 7.5 Intelligent Construction and Its Related Technologies

fields. The integration of related technologies is also on this route. Therefore, along the line of "perception, analysis, decision-making, and execution", we can roughly sort out the various technologies required for intelligent construction. Figure 7.5 shows the intelligent construction and its related technologies. It is currently only a preliminary summary. With the development of technology, there will be more integration of new technologies.

The common technology is "perception, analysis, decision-making." The "perception" is the perception terminal, which involves data collection technology. "Analysis and decision-making" is the function of the computer. It is the core of AI that includes machine learning algorithms. The "execution" involves extensive knowledge of various fields, but the control technology has certain commonalities. Behind these techniques, there is the same theoretical basis, as shown in Fig. 7.6.

It can be found from Fig. 7.6 that the knowledge involved in intelligent construction finally points to physics, mathematics, and chemistry. This is not surprising because the foundation of modern science and technology (characterized by traditional civil, mechanical, and electrical industries) is mechanics theory, which belongs to classical physics. In contrast, the foundation of modern science and technology (characterized by information technology) is an electromagnetic theory, which belongs to modern physics study. Mathematics is a tool of physics, and the two complement each other.

The role of transportation infrastructure (structure) is to support the operation of vehicles (cars, trains, planes) and ensure safety and comfort. Therefore, whether it is a highway, a railway, an airport, or an urban road, it is essentially a problem of mechanics. Materials are the basic elements that make up a structure, and there will be more and more types of materials involved, including smart materials in the future. For organic materials (asphalt mixture), it is essentially a chemical problem, while other materials are mostly related to physics (material structure). This is why the core knowledge of intelligent construction is attributed to mathematics, physics, and chemistry, and it is also the knowledge that must be possessed by primitive innovation, but it is not classical physics based on mechanics, but modern physics.

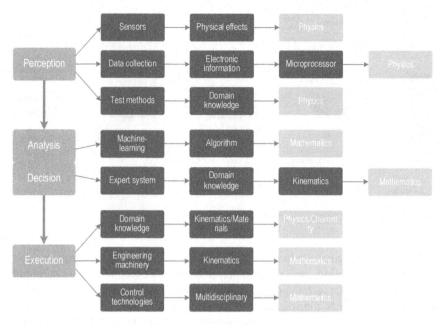

Fig. 7.6 Core knowledge of intelligent construction

In order to improve the quality of engineering and finally build intelligent transportation infrastructure, it is not enough to rely on traditional domain (industry) knowledge. The relevant technologies included in intelligent construction have been given above, and the key is how to master them. For technicians in the civil engineering industry, electronic information technology (especially hardware) is the most lacking knowledge and is generally considered the most difficult knowledge to master. In fact, it is not difficult. You can master it according to the black box method. The following takes the sensing device as an example for a brief description.

The scope of electronic and information technology is extensive. Civil engineers and practitioners need only master the application sides of the technology to be practical. The electronic information technology in ICT includes sensing devices with the embedded microcomputer. The microcomputer is also known as a controller MCU. MCU's essential element is the microprocessor (MPU). Microprocessors are ubiquitous in household appliances, automobile control, engineering machinery control, testing equipment, etc. The microprocessor can form a sensing and control device by combining other circuits as a critical component of various intelligent devices.

The fundamental knowledge of a microprocessor chip includes inputs, outputs, and functions. ICT developers need to control each chip pin by writing a program to perform various operations, such as data acquisition and mechanical control. Then, such a chip would often work in conjunction with other circuits and components. Please refer to Volume two and three for further details on electronic information technology.

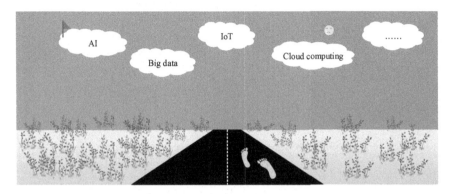

Fig. 7.7 The foundation is the most important

Emerging technologies are involved in intelligent construction, as shown in Fig. 7.4. Fundamental science and technology are still the keys.

Intelligent construction is the product of multi-disciplinary crossover, but the division between disciplines is artificial. Many theories and technologies are essentially the same, but the terms may differ in different disciplines. The basic knowledge has not changed substantially, and the basics are still the original ones. It is equivalent to a castle in the air without a solid foundation. No matter how many new terms, it is only a different way of expression, and there is no fundamental innovation. Rather than immersing yourself in the fun of learning new terms, it's better to get down-to-earth and master the fundamental skills.

Intelligent construction has brought new hope to the traditional construction industry and will accelerate the pace of innovation in the civil engineering industry. Seize the opportunity of intelligent construction, consolidate the foundation (Fig. 7.7), and expand the knowledge structure. This is the greatest wish of this book!

References

Brindley, Keith. Starting Electronics. Newnes, 1999.

Lucci, Stephen. Artificial Intelligence in the 21st Century. Mercury Learning and Information, 2015.

Mayer-Schonberger, Viktor, and Cukier, Kenneth. Big Data - A Revolution That Will Transform How We Live, Work, and Think. HarperCollins Publisher, 2014.

Negnevitsky, Michael. Artificial intelligence - a guide to intelligent systems, Pearson Education Canada, 2011.

Scaruffi, Piero, The Nature of Consciousness. Omniware, 2006.

Strategic Advisory Committee of China Manufacturing Power Construction, Strategic Advisory Center of Chinese Academy of Engineering. Made in China 2025 Series, Intelligent Manufacturing, Beijing: Publishing House of Electronics Industry, 2016. [in Chinese]

Sutton, Richard S, and Barto, Andrew G. Reinforcement Learning - An Introduction. A Bradford Book, 2018.

Wang, Zhiliang, Internet of Things Now and Future, China Machine Press, 2010. [in Chinese]

Xu, Guozhi. System Science. Shanghai Science and Technology Education Publishing House, 2001. [in Chinese]

Xu, Guanghui. Dynamic Principle and Engineering Application of Subgrade Continuous Compaction Control. Science Press, 2016. [in Chinese]

Xu Guanghui. High-speed railway subgrade continuous and intelligent compaction control technology. China Railway Publishing House, 2019. [in Chinese]

Yongquan, Liangxing and Yang, Ruilong. The mathematics of deep learning. People's Posts and Telecommunications Press, 2019. [in Chinese]

© China Railway Publishing House Co., Ltd. 2023   227
G. Xu and D. Wang, *Introduction to Intelligent Construction Technology of Transportation Infrastructure*, Springer Tracts in Civil Engineering,
https://doi.org/10.1007/978-3-031-13433-3

Printed in the United States
by Baker & Taylor Publisher Services